THE BUG CREEK PROBLEM AND
THE CRETACEOUS-TERTIARY TRANSITION
AT MCGUIRE CREEK, MONTANA

Jinx,
I hope you like
the books, its
5 years of
my life!
Cheers,
Dad

D1490236

# The Bug Creek Problem and the Cretaceous-Tertiary Transition at McGuire Creek, Montana

Donald L. Lofgren

UNIVERSITY OF CALIFORNIA PRESS
Berkeley • Los Angeles • London

UNIVERSITY OF CALIFORNIA PUBLICATIONS IN GEOLOGICAL SCIENCES

Editorial Board: Stanley M. Awramik, Anthony D. Barnovsky,
James A. Doyle, Mary L. Droser, Peter M. Sadler

Volume 140
Issue Date: May 1995

UNIVERSITY OF CALIFORNIA PRESS
BERKELEY AND LOS ANGELES, CALIFORNIA

UNIVERSITY OF CALIFORNIA PRESS, LTD.
LONDON, ENGLAND

Library of Congress Cataloging-in-Publication Data

Lofgren, Donald L., 1950–
    The Bug Creek problem and the Cretaceous-Tertiary
transition at McGuire Creek, Montana / Donald L. Lofgren.
        p.    cm.  — (University of California publications in
geological sciences; v. 140)
    Includes bibliographical references (p.    -    ).
    ISBN 0-520-09800-5 (pbk.)
    1. Vertebrates, Fossil—Montana—McGuire Creek Region
(McCone County)  2. Cretaceous-Tertiary boundary—
Montana—McGuire Creek Region (McCone County)
3. Paleontology, Stratigraphic.    I. Title.   II. Series.
QE734.5.L64   1995
566'.09786'26—dc20                                    95-12046
                                                              CIP

The paper used in this publication meets the minimum
requirements of American National Standard for Information
Sciences—Permanence of Paper for Printed Library Materials,
ANSI Z39.48-1984.

# Contents

# Acknowledgments

This research effort was part of a continuing University of California Museum of Paleontology (UCMP) field program headed by Dr. Wiliam A. Clemens. I thank those associated with the UCMP who helped to see this project to completion, most notably Dr. Clemens who suggested and supervised the project, and the members of UCMP field parties during the years 1984 to 1989 especially: Mark Goodwin, Dr. J.H. Hutchison, Kyoko Kishi, Dr. Laurie Bryant, Allen Tedrow, Matthew Colbert, James Foster, Dr. Anthony Runkel, and Dr. Zhexi Luo. Special thanks to Mark Goodwin and Kyoko Kishi for fossil preparation and to those who sorted screenwash concentrate from McGuire Creek.

I am grateful to the many local ranchers who allowed access to their land, and for their friendship and hospitality, including Lester and Cora Engdahl, Robert and Jane Engdahl, Judd and Jay Twitchell, Clay Taylor, and their families. Don and Marge Beckman of Fort Peck contributed the use of a motorboat which made fieldwork and fossil collection in areas far from roads logistically feasible.

For their helpful comments and discussions I thank, Dr. Bryant, Dr. Runkel, Dr. J. David Archibald, Dr. Carol Hotton, Dr. Douglas Nichols, Dr. J. Keith Rigby Jr., Dr. Carl Swisher, Dr. Lowell Dingus, Dr. Jennifer Hogler, Dr. Zhexi Luo, Dr. Judd Case, Dr. Richard Fox, Gayle Nelms, and Robert Dundas. Special thanks to Dr. Hutchison for hours of stimulating discussion on various aspects of the Cretaceous-Tertiary transition in eastern Montana.

Dr. Archibald, Dr. J.A. Lillegraven, and Dr. P. Sadler carefully reviewed the original manuscript and provided many helpful suggestions. Dr. Hotton kindly allowed me to report the age of a number of rock samples from McGuire Creek based on her analyses and interpretations of palynofloras.

Specimens were loaned by Dr. L. Van Valen of the University of Chicago and Mary Ann Turner of the Yale Peabody Museum. Financial support was provided by the Department and Museum of Paleontology, University of California at Berkeley (Annie M. Alexander Endowment); the University of California at Berkeley (Regents Fellowship and the Alfred F. Moore and Chella D. Moore Scholarship); National Science Foundation Grant no. BSR-85-13253 to Dr. Clemens; and the Board of Trustees of the Raymond M. Alf Museum.

I especially want to thank my parents Lionel and Delores Lofgren, and my wife Laura, for their support during my academic studies, which made completion of this project possible.

# Abbreviations

Institutions

| | |
|---|---|
| AMNH | American Museum of Natural History, New York |
| CM | Carnegie Museum of Natural History, Pittsburgh |
| KU | Museum of Natural History, University of Kansas, Lawrence |
| PU | Princeton University, Princeton, New Jersey |
| SPSM | St. Paul Science Museum, Minnesota |
| UA | University of Alberta |
| UCM | University of Colorado Museum, Boulder |
| UCMP | University of California Museum of Paleontology, Berkeley |
| UMVP | University of Minnesota Museum of Paleontology, Minneapolis |
| USNM | United States National Museum, Washington, D.C. |
| YPM | Yale Peabody Museum, New Haven, Connecticut |

Dental Measurements
(See Archibald 1982: fig. 1 and pp. xv-xvi, for more complete definitions of dental measurement abbreviations.)

| | |
|---|---|
| L | Length |
| W | Width |
| W-A | Anterior Width |
| W-P | Posterior Width |
| L/W | Length relative to Width |
| L/W-A | Length relative to Anterior Width |
| L/W-P | Length relative to Posterior Width |
| W-Tri | Trigonid Width |
| W-Tal | Talonid Width |
| W-Tri/W-Tal | Trigonid Width relative to Talonid Width |
| L/W-Tri | Length relative to Trigonid Width |
| L/W-Tal | Length relative to Talonid Width |
| A | Distance between Apex and Lingualmost Point of Protocone |
| A/W-A | A (defined above) relative to Anterior Width |

Localities

| | |
|---|---|
| BA | Bug Creek Anthills |
| BG | Brown-Grey Channel |
| BS | Black Spring Coulee |
| BW | Bug Creek West |
| CBB | Chris' Bonebed |
| F1 | Frenchman 1 |
| FC | Flat Creek 5 |
| FR | Ferguson Ranch |
| HE | Hell's Hollow |
| HH | Harbicht Hill |
| JC | Jacks Channel |
| KM | K-Mark |
| LF | Long Fall |
| LR | Little Roundtop |
| M | Mantua Lentil |
| MD | Matt's Dino Quarry |
| SL | Second Level |
| SR | Shiprock |
| UU | Up-Up-the-Creek |
| WC | Worm Coulee 1 |
| WK | Wounded Knee |
| ZL | Z-Line |

Miscellaneous

| | |
|---|---|
| bk | "Bugcreekian" North American Land Mammal "Age" |
| IHS | Inclined heterolithic stratification |
| MCZ | McGuire Creek "Z" Lignite |
| MNI | Minimum number of individuals |
| NALMA | North American Land Mammal "Age" |
| NISP | Number of identifiable specimens per taxon |
| Pu | Puercan North American Land Mammal "Age" |
| TL | Tonstein Lignite |

# Abstract

The "Bug Creek Problem" refers to questions concerning the nonmarine vertebrate biostratigraphy of the upper Hell Creek and lower Tullock formations, an interval where significant faunal turnover is evident. This stratigraphic interval, which spans the Cretaceous-Tertiary transition, contains three distinct types of vertebrate assemblages, Lancian-Bug Creek-Puercan, and perhaps the youngest known records of dinosaurs in North America. Bug Creek assemblages are faunally intermediate to Lancian (Cretaceous mammals/dinosaurs) and Puercan (Paleocene mammals/no dinosaurs) assemblages, which suggests that faunal turnover during the Cretaceous-Tertiary transition was not abrupt. However, several issues concerning Bug Creek assemblages are in dispute, including: (1) their geochronologic age(s); (2) whether they contain reworked fossils; and (3) whether they are temporally intermediate between Lancian (Cretaceous) and Puercan (Paleocene) faunas. Bug Creek assemblages are critical to Cretaceous-Tertiary extinction debates because they can be interpreted to support both gradual and catastrophic models of faunal turnover during the Cretaceous-Tertiary transition.

Lithofacies and biostratigraphic (including pollen) analyses of the upper Hell Creek and lower Tullock formations in the vicinity of McGuire Creek, McCone County, Montana, were employed to address contentious issues surrounding Bug Creek assemblages. Seven lithostratigraphic facies were recognized at McGuire Creek, and analysis of facies indicates that Bug Creek assemblages are restricted to lag deposits of large channel fills in the upper Hell Creek Formation which are deeply entrenched into floodplain deposits that yield in situ dinosaur remains. A locally traceable erosion surface was created by these channeling events. Because an enormous amount of older sediment was removed during channel entrenchment, the probability that Bug Creek assemblages contain reworked fossils is very high. Additional evidence indicating that the dinosaurian and Lancian mammal components of Bug Creek assemblages are reworked is: (1) Bug Creek assemblages are present only in channel fills that were entrenched into Lancian (Cretaceous) strata; and (2) when traced laterally, channel facies that contain Bug Creek assemblages (with dinosaurs) interfinger with floodplain deposits that lack dinosaurs.

Complete channel-fill sequences in the upper Hell Creek Formation are overlain by a laterally extensive lignite, the McGuire Creek "Z" (MCZ) (base of the Tullock Formation), which is overlain by pond deposits.

Palynology was employed to distinguish Cretaceous and Paleocene strata at McGuire Creek. All channel fills containing Bug Creek assemblages that could be sampled for pollen yielded Paleocene palynofloras. Thus, the erosion surface created by the entrenchment of channels yielding Bug Creek assemblages into floodplain deposits with in situ dinosaurs is the delineation of the local Cretaceous-Tertiary boundary in a large part of the McGuire Creek area.

Palynology yields further evidence that reworking has occurred, because: (1) channel fills with Bug Creek assemblages are palynologically Paleocene, while floodplains with the highest in situ dinosaur remains yield Cretaceous palynofloras; (2) the lowest Paleocene floodplain deposits (identified palynologically) lack dinosaurs; and (3) dinosaur remains encased in mudstone that yielded a Cretaceous palynoflora are present in a Paleocene channel fill at Black Spring Coulee (see Lofgren et al., 1990). Thus, data from McGuire Creek indicate that the presence of dinosaurs in Paleocene channels is the result of reworking, not of their survival into the Paleocene.

Development of a biostratigraphic zonation for the upper Hell Creek Formation at McGuire Creek is limited because of complex channeling, and channel fills can be temporally ordered only in rare instances. Thus, biochronologic criteria form the main basis for temporally ordering the three types of assemblages as Lancian-Bug Creek-Puercan (oldest to youngest).

McGuire Creek local faunas are referred to the Lancian North American Land Mammal "Age" (NALMA) and the Pu0 and Pu1 interval zones of the Puercan NALMA (Archibald and Lofgren, 1990). Analysis of the informal subdivision of the Pu0 interval zone into biochrons (bk1-bk2-bk3) indicates that the type localities (Bug Creek Anthills-bk1; Bug Creek West-bk2;, Harbicht Hill-bk3) require additional sampling and new systematic treatments before these biochrons can be abandoned or formally adopted.

The survival rates of the dinosaurian and mammalian part of the latest Cretaceous vertebrate fauna may or may not be compatible with catastrophic terminal Cretaceous extinction, depending on the provenance of the Lancian component of Bug Creek assemblages. At McGuire Creek, the survival rates of mammal species (approximately 10%) and dinosaur genera (0%) are low across the Cretaceous-Tertiary boundary because the presence of dinosaurs and certain Lancian mammals in Bug Creek assemblages (palynologically Paleocene) are interpreted to be the result of reworking.

# INTRODUCTION

The "Bug Creek Problem" refers to a number of unresolved issues concerning the Bug Creek vertebrate assemblages and nonmarine biostratigraphy of the uppermost Hell Creek and lower Tullock formations of eastern Montana. The mammalian component of Bug Creek assemblages is intermediate in composition between typical Late Cretaceous and Early Paleocene faunas, but detailed systematic and biostratigraphic study of these assemblages has never been completed. Also, it has not been determined whether the presence of dinosaurs and certain Cretaceous mammals in Bug Creek assemblages is a taphonomic artifact caused by reworking. This lack of data has fueled the debate concerning patterns of faunal turnover of the terrestrial biota during the Cretaceous-Tertiary transition. These uncertainties are illustrated by the fact that the Bug Creek faunal data have been used to support both catastrophic (Smit and Van der Kaars, 1984; Smit, 1985; Smit et al., 1987) and noncatastrophic models (Clemens and Archibald, 1980; Archibald, 1981, 1982, 1984, 1987c; Clemens, 1983; Archibald and Clemens, 1984) of vertebrate turnover during the K-T transition in eastern Montana.

Unresolved questions can be divided into four main issues:
(1) Do Bug Creek vertebrate assemblages represent the remains of animals that lived contemporaneously or were Cretaceous fossils eroded and redeposited with Paleocene vertebrate remains?
(2) Based on biostratigraphic and/or biochronologic criteria, are Bug Creek assemblages temporally intermediate between Lancian (Late Cretaceous) and Puercan (Early Paleocene) assemblages?
(3) Can Bug Creek assemblages themselves be placed in a temporal sequence that reflects the faunal differences between individual local faunas?
(4) What are the geochronologic age(s) of Bug Creek assemblages?

## GEOLOGIC AND BIOSTRATIGRAPHIC SETTING

The first paleontological-geological exploration of Garfield and McCone counties, Montana (Figures 1a-c), was undertaken by Barnum Brown of the American Museum of Natural History, who described the dinosaur-bearing "Hell Creek beds" (Brown, 1907). Later, Rogers and Lee (1923) and Thom and Dobbin (1924) recognized the Hell

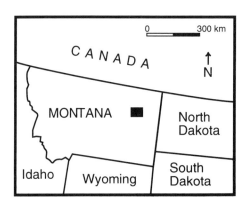

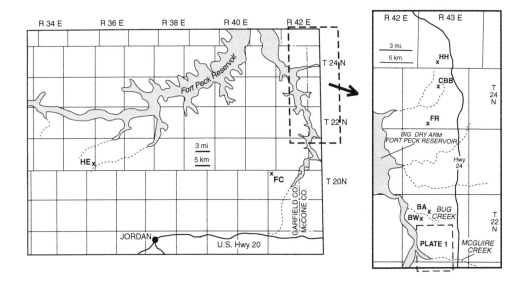

FIGURE 1a. Location of study area in northeastern Montana.

FIGURE 1b. Location of sites of paleontological interest in Garfield County, Montana. Fossil localities/local faunas are: (HE) Hell's Hollow, (FC) Flat Creek.

FIGURE 1c. Location of sites of paleontological interest in McCone County, Montana. Fossil localities/local faunas are: (HH) Harbicht Hill, (CBB) Chris' Bone Bed, (FR) Ferguson Ranch, (BA) Bug Creek Anthills, (BW) Bug Creek West. Heavy dashed lines outline the McGuire Creek study area. Plate 1 is a partial geologic map of the McGuire Creek area.

Creek "beds" and the Tullock "Formation" as members of the Lance Formation. Wilmarth (1938) elevated both to the rank of formation. Roland Brown (1952) proposed a formal lithostratigraphic boundary between the Hell Creek and Tullock formations at the base of the first laterally persistent lignite bed above the highest occurrence of dinosaurs. This lignite, termed the "Z" coal bed in McCone County by Collier and Knechtel (1939), was used by Roland Brown (1952) to approximate the local Cretaceous-Tertiary boundary, assuming that the passing of the last dinosaur signaled the close of the Cretaceous.

Paleontological research in eastern Montana was minimal for five decades following Barnum Brown's (1907) early work. In the meantime, North American vertebrate paleontologists developed a series of provincial time units based on nonmarine mammal-bearing rock units that encompassed the latter part of the Late Cretaceous and all Cenozoic epochs. Of interest here are those units which occur near the Cretaceous-Tertiary boundary: the Lancian, "Bugcreekian," and Puercan North American Land Mammal "Ages." North American Land Mammal "Ages" (abbreviated to NALMA) are not true ages, because in most cases they are based on aggregates of mammal genera, not on corresponding chronostratigraphic standards (see Lillegraven and McKenna, 1986, for detailed discussion of NALMAs). Savage (1962) advocated the placing of "Ages" in quotation marks to denote this fundamental difference between stage-ages and NALMAs.

The Puercan NALMA was proposed in 1941 by the Wood Committee (Wood et al., 1941), based on the mammalian fauna from the Puerco Formation (now lower part of the Nacimiento Formation), San Juan Basin, New Mexico. The Puercan NALMA was considered to be Paleocene in age. Dorf (1942) proposed the Lancian "Age" based on the Lance Formation at its type locality, Lance Creek, Wyoming, and suggested that the Lancian "Age" could be recognized by mammals and dinosaurs of the "*Triceratops* zone" and plants from the Lance Formation. The Lancian NALMA was considered to be Late Cretaceous in age (Wood et al. 1941). This was the status of North American nonmarine Cretaceous-Tertiary temporal units at the time of the Bug Creek discoveries.

In the early 1960's, several fossil-rich vertebrate sites from channel fills in the Hell Creek and Tullock formations in the vicinity of Bug Creek, McCone County, Montana (Figure 1c), were discovered and briefly described by R.E. Sloan and L. Van Valen (1965; Van Valen and Sloan, 1965). Ken's Saddle, a locality from the upper Hell Creek Formation was described as being of typical Late Cretaceous aspect (or Lancian in age) in that it contained a fauna similar to that from the Lance Formation (dinosaur remains and a Lancian mammalian assemblage). A Puercan mammal fauna, Purgatory Hill, from the lower Tullock Formation was also mentioned (Sloan and Van Valen, 1965) and briefly described (Van Valen and Sloan, 1965). Of special interest were three channel fillings in the uppermost Hell Creek Formation which yielded dinosaurs and mammals of Late Cretaceous aspect (Lancian), along with mammals whose descendants are characteristic of the Puercan (Sloan and Van Valen, 1965). The three sites, Bug Creek Anthills, Bug Creek West, and Harbicht Hill (Figure 1c), which were referred to as the Bug Creek Faunas, contained the first records of the multituberculates *Catopsalis* (first

Table 1. Vertebrate assemblages near the Cretaceous-Tertiary boundary in McCone County, Montana[a]

| GEOLOGIC AGE | LANCIAN ASSEMBLAGE LATE CRETACEOUS | BUG CREEK ASSEMBLAGE[b] AGE UNCERTAIN | PUERCAN ASSEMBLAGE EARLY PALEOCENE |
|---|---|---|---|
| **MULTITUBERCULATA** | | | |
| *Meniscoessus* | XX | XX | |
| *Essonodon* | XX | XX | |
| *Cimolomys* | XX | XX | |
| *Cimolodon* | XX | XX | |
| *Mesodma* | XX | XX | XX |
| *Cimexomys* | XX | XX | XX |
| *Stygimys* | | XX | XX |
| *Catopsalis* | | XX | XX |
| **MARSUPIALIA** | | | |
| *Alphadon* | XX | XX | |
| *Pediomys* | XX | XX | |
| *Didelphodon* | XX | XX | |
| *Glasbius* | XX | XX | |
| *Peradectes* | | XX | XX |
| **EUTHERIA** | | | |
| *Gypsonictops* | XX | XX | |
| *Batodon* | XX | XX | |
| *Cimolestes* | XX | XX | XX |
| *Procerberus* | | XX | XX |
| *Protungulatum* | | XX | XX |
| *Mimatuta* | | XX | XX |
| *Oxyprimus* | | XX | XX |
| *Ragnarok* | | XX | XX |
| **DINOSAUR REMAINS** | XX | XX | |

a. Only generic-level mammalian data and presence or absence of dinosaurian remains are given.

b. The original Bug Creek assemblages are included as the initial interval zone (Pu0) in the Puercan NALMA by Archibald and Lofgren (1990). This has no effect on the uncertainty of their geochronologic age assignment.

North American record) and *Stygimys*, the "proteutherian" *Procerberus*, and the archaic eutherian ungulates ("condylarths") *Protungulatum*, *Mimatuta*, *Oxyprimus*, and *Ragnarok* (Table 1). Based on their mammalian assemblages and stratigraphic position, the Bug Creek Faunas (here designated Bug Creek assemblages) were inferred by Sloan and Van Valen (1965) to be younger (from a higher horizon) than the Lancian fauna, Ken's Saddle, but were considered to have been deposited before the early Puercan ("*Ectoconus* zone"). Therefore, the Bug Creek assemblages were intermediate in composition between previously known Lancian/Late Cretaceous and Puercan/Paleocene assemblages (such as those present at Ken's Saddle and Purgatory Hill) and it was implied that the Bug Creek assemblages represented a previously unsampled faunal interval situated between the Lancian and Puercan NALMAs (Sloan and Van Valen, 1965; Van Valen and Sloan, 1965). Sloan and Van Valen (1965) considered the Bug Creek assemblages to be Cretaceous in age, using two criteria: (1) the Bug Creek sites were stratigraphically below the coal bed (termed the "Z" coal) used to approximate the position of the local K-T boundary in Roland Brown's (1952) formula of first coal above the highest occurrence of dinosaur remains ("Z" coal was also used as the lithostratigraphic boundary between the Hell Creek and Tullock formations); (2) dinosaur remains were present in the Bug Creek assemblages.

Using the stratigraphic distance of each locality above or below the "Z" coal, Sloan and Van Valen (1965) placed the five sites in the following temporal sequence, from oldest to youngest: Ken's Saddle, Bug Creek Anthills, Bug Creek West, Harbicht Hill, and Purgatory Hill. With the five sites sequentially ordered in this fashion, Van Valen and Sloan (1977) argued that the typical Late Cretaceous vertebrate assemblage (or *Triceratops* community) was gradually replaced by an assemblage with Paleocene affinities (or *Protungulatum* community) and that the turnover began stratigraphically below the "Z" coal or during the latest Cretaceous.

Archibald (1982) proposed a model of vertebrate turnover in eastern Montana similar to Van Valen and Sloan's (1977), whereby two faunal-facies were present in the upper Hell Creek Formation prior to the close of the Cretaceous: a Hell Creek faunal-facies (the *Triceratops* community of Van Valen and Sloan, 1977) from siltstones, and a Bug Creek faunal-facies (the *Protungulatum* community of Van Valen and Sloan, 1977) from large sandstone channel fills (Archibald, 1981, 1982). From their stratigraphic position the two faunal-facies were interpreted to be in part coeval, but were separated by ecological differences. Limited magnetostratigraphic data indicated that the two faunal-facies occurred during the same period of reversed polarity (Archibald et al., 1982). Archibald (1981, 1982) concluded that prior to the close of the Cretaceous, faunal turnover occurred gradually within the Bug Creek faunal-facies concurrent with a rapid turnover within the Hell Creek faunal-facies.

Testing the proposal of an extraterrestrial cause for catastrophic, terminal Cretaceous extinctions (Alvarez et al., 1980; Alvarez, 1986) required greater precision in stratigraphic studies of the Bug Creek assemblages. The Bug Creek faunal data were in apparent conflict with a catastrophic extinction hypothesis because the changeover between Late Cretaceous (*Triceratops* community, or Hell Creek faunal-facies) and

Early Paleocene faunas (*Protungulatum* community, or Bug Creek faunal-facies) was evidently in progress before the close of the Cretaceous (Sloan, 1976; Van Valen and Sloan, 1977; Clemens et al., 1981; Archibald, 1982; Van Valen 1984).

An iridium enrichment, which is presumed to be the record of bolide impact (Alvarez et al., 1980), occurs within the "lower Z" coal (approximate K-T boundary) in western Garfield County (Smit and Van der Kaars, 1984). A number of typically Cretaceous palynoflora species are not found above the level of iridium enrichment when it is present in a local section (Hotton, 1984, 1988). Also, new appearances of Paleocene pollen species are rare in the first few meters above the iridium enrichment. Therefore, Paleocene palynofloras are distinguished from Cretaceous palynofloras by the extinction of Cretaceous species (Hotton, 1988). Because an iridium enrichment was not found in McCone County within the stratigraphic level comparable to that of the "lower Z" coal of Garfield County, an absence of Cretaceous pollen species was employed to determine the local K-T boundary in McCone County (in Montana, Bug Creek assemblages are known only from McCone County). Harbicht Hill, Bug Creek West, and a recently discovered Bug Creek site at Ferguson Ranch (Figure 1c) yielded palynofloras characterized by the absence of typical Cretaceous species and were therefore considered to be Paleocene in age (Smit and Van der Kaars, 1984; Smit, 1985; Sloan et al., 1986; Smit et al., 1987).

Subsequent debates on whether the Bug Creek faunal sequence began in the Cretaceous were centered on the Bug Creek Anthills locality. Bug Creek Anthills was argued to be Cretaceous (Archibald, 1984; Sloan et al., 1986; Rigby et al., 1987; Rigby, 1987; Newmann, 1988); Paleocene (Smit and Van der Kaars, 1984; Smit, 1985; Retallack et al., 1987); or of indeterminable age (Dingus, 1984; Fastovsky and Dott, 1986; Fastovsky, 1987; Smit et al., 1987). Detailed study of the geology in the vicinity of the Bug Creek Anthills locality indicated that correlation of this site is uncertain because of repeated episodes of incision and filling of fluvial channels (Fastovsky and Dott, 1986).

Because the Bug Creek assemblages were restricted to large sandy channel fillings, these assemblages were viewed by some workers as Paleocene, but containing reworked Cretaceous fossils (Smit and Van der Kaars, 1984; Bryant et al. 1986; Retallack et al., 1987). In contrast, others viewed dinosaur remains in these same channel fills as evidence of dinosaur survival into the Paleocene (Rigby, 1985, 1987; Rigby and Sloan, 1985; Sloan et al., 1986; Rigby, et al., 1987). However, Fastovsky (1987) concluded that provenance of the Bug Creek assemblages is uncertain because of taphonomic considerations imposed by the depositional setting of these vertebrate remains. Also, based on analyses of the sedimentary record spanning the K-T transition in eastern Montana, both Dingus (1983, 1984) and Fastovsky (1986, 1987) concluded that the chronostratigraphic resolution of these sediments is not sufficiently refined to resolve the K-T debates.

Therefore, while the age, provenance, and biostratigraphy of the original Bug Creek assemblages are central to K-T boundary debates, data from the original Bug Creek sites are not sufficient to resolve the issues being debated.

## MCGUIRE CREEK STUDY AREA

I conducted geologic and biostratigraphic analyses of the upper Hell Creek and lower Tullock formations in the vicinity of McGuire Creek because the area contains abundant Bug Creek-like vertebrate assemblages and extensive exposures of the stratigraphic interval in which the Cretaceous-Tertiary and Lancian-Puercan transitions occur. Thus, abundant faunal data with potential for good stratigraphic control made the McGuire Creek area ideally suited to address contentious issues concerning the Bug Creek assemblages. Results garnered from this study are divided into six parts:

1. *Geological and Palynological Correlations*: Using facies analysis and pollen analysis, I determined that channel facies with Bug Creek assemblages do not correlate to floodplain facies that yield in situ dinosaur remains. Also, I constructed a sequence of geologic events spanning the K-T transition at McGuire Creek which shows that developing a local biostratigraphic zonation for the upper Hell Creek Formation is extremely difficult because of multiple episodes of channel entrenchment.

2. *Reworking of Fossils*: I concluded that Bug Creek assemblages certainly contain reworked fossils, and that the survival of dinosaurs into the Paleocene is a taphonomic artifact caused by reworking.

3. *McGuire Creek Biochronology and the "Bugcreekian Age"*: From the results from 1 and 2 above, I concluded that current mammalian zonations spanning the K-T transition are problematic, but were developed on the best biochronologic data available.

4. *Cretaceous-Tertiary Correlations*: I determined that palynology is the best method to employ to recognize the terrestrial K-T boundary at McGuire Creek and that all Bug Creek assemblages for which data is available yield Paleocene pollen.

5. *Faunal Turnover During the K-T Transition in Montana*: I examined mammalian survivorship across the K-T boundary and concluded that survivorship rates vary greatly, depending on how the presence of Lancian mammals in Bug Creek assemblages and the age of Bug Creek assemblages are interpreted.

6. *Appendices 1 and 2, on Mammalian Systematics and Faunal Data*: I analyzed the systematic paleontology of McGuire Creek mammals (and a few other sites) and listed mammalian taxa present at each McGuire Creek locality to provide a biostratigraphic database for this study.

# MATERIALS AND METHODS

Geological and palynological investigations in the McGuire Creek area were limited to the stratigraphic interval of the upper 40 m of the Hell Creek Formation and the lower 20 m of the Tullock Formation. Field data were plotted directly onto standard USGS 7.5-minute quadrangles (Bug Creek and Nelson Creek Bay) that were enlarged 200% and later transferred to a 200% enlarged base map (Plate 1). Section thicknesses were measured in meters, using a hand level and jacob's staff.

Seven lithofacies (A-G) were recognized and mapped and the stratigraphic position of each is indicated in cross sections (Plates 2-4) and on a partial geologic map (Plate 1). Facies mapping was not completed in the northern and extreme southwestern part of the study area. For clarity in presentation, only the three channel-fill facies (B, E, F) and the coal-swamp facies (C) are indicated on the geologic map (Plate 1). Facies C lignites were used to correlate measured sections. Recognition and mapping of individual channel-filling events were completed where possible. Certain channel fills are traceable through laterally continuous exposures, while others are recognizable across disjunct exposures based on lithologic similarity and congruence of paleocurrent azimuths measured from crossbedded sandstones. The remainder of channel-fill exposures are mapped as undifferentiated, because they could not be correlated with the recognized channel fills. Although three types of vertebrate assemblages are found in channel fills at McGuire Creek (Lancian, Bug Creek, and Puercan; Table 1), faunal content was never employed to aid in recognition of individual channel fills.

Several techniques of fossil collection were employed at McGuire Creek, and the specific technique(s) used for each locality are listed in Appendix 2. All vertebrate sites were prospected for fossils by visual scanning of the available surface exposure. Hand quarrying was carried out at a few sites where recovery of articulated or associated skeletal material was desirable. Screenwashing techniques, described by McKenna (1965), were employed at many sites. Isolated teeth collected from mounds constructed by harvester ants were used to a limited degree. The ability of ants to concentrate fossils is well documented (Hatcher, 1896; Galbreath, 1959; Clemens, 1964; Clark et al., 1967; Macdonald, 1972), and anthill concentrates are useful for adding qualitative knowledge of the small vertebrate fauna (Clark et al., 1967). But anthill concentrations are problematic, because harvester ants can mix fossils from different stratigraphic units. Transport experiments using glass or plastic beads indicate that ants can carry anthill material a minimum of 50 feet (Clemens, 1964). Also, anthill collections are biased toward small vertebrate elements because of the size selectivity of ants. Because of these biases, anthill collections were not included in quantitative analyses. Localities that include fossils from anthills are indicated in Appendix 2.

The University of California Museum of Paleontology (UCMP) records each vertebrate locality by using a V followed by numbers (e.g., V87030). Specimens are given a numerical code preceded by UCMP (e,g., UCMP 113465).

Unweathered rock samples analyzed for pollen and spore content were collected by digging beyond the weathered outcrop surface. Rock samples were prepared by standard palynological techniques and sent for analysis to Dr. Carol Hotton, who made a Cretaceous or Paleocene age assignment for each sample, based on comparison to stratigraphically controlled samples from Garfield County, Montana (Hotton, 1988).

Specific definitions of two terms, "Bug Creek vertebrate assemblage" and "local fauna" are required for clarity in later discussions. A Bug Creek vertebrate assemblage is defined as an aggregation of vertebrate fossils containing a mixture of Lancian or latest Cretaceous mammals and dinosaurs with mammals more typically indicative of a Puercan or Paleocene age (Table 1). Therefore, the term "Bug Creek vertebrate assemblage", or "Bug Creek assemblage" for short, refers to the aggregation of taxa listed in the middle column of Table 1, wherein mammalian taxa from both the left (Lancian) and right (Puercan) columns are present in some combination, along with dinosaur remains.

The term "local fauna" is employed at McGuire Creek in a manner similar to that advocated by Tedford (1970), but additional comments are necessary. As employed at McGuire Creek, vertebrate fossils within lag concentrations from individually mapped channel fills constitute a local fauna. Therefore, at McGuire Creek, local faunas are aggregates of fossil species collected from a restricted stratigraphic interval that is local in time and space (a distinct channel-filling event). However, fossil species collected from this depositional setting represent a thanatocoenose, rather than a biocoenose. Therefore, the term local fauna is used to describe the fossil species present in one channel fill, whether or not these animals actually coexisted.

A detailed taxonomic database is necessary for accurate assessment of biostratigraphic correlations and is especially critical when assessing the faunal turnover at the Cretaceous-Tertiary boundary. Therefore, a thorough but not exhaustive systematic review of the McGuire Creek mammalian faunas is given in Appendix 1. In many instances it was necessary to study specimens and epoxy casts in the UCMP collections from Bug Creek Anthills, Bug Creek West, and Harbicht Hill, in the upper Hell Creek Formation, Montana, and Mantua Lentil, in the Polecat Bench (Fort Union) Formation, Wyoming, to aid in identification of McGuire Creek species. Hence, some faunal data from these sites are included in this study.

The mammalian classification scheme used here follows that of Marshall et al. (1990) for marsupials, Novacek (1986) for eutherians, and Hahn and Hahn (1983) for multituberculates.

Works that involve Late Cretaceous-Early Tertiary mammals by Archibald (1982),

Clemens (1964, 1966, 1973), Sloan and Van Valen (1965), Lillegraven (1969), and Van Valen (1978) were followed where applicable. Modifications and additions to the morphological descriptions of taxa presented in these studies are given where appropriate, based on specimens recovered from McGuire Creek. No new species are named.

Only dental remains were studied. Isolated elements and fragments of postcranial material were recovered but are not included in this study.

Cusp terminology for therian teeth is that of Van Valen (1966: figure 1). Stylar cusps of marsupials are named in the manner outlined by Clemens (1966). Measurements of therian molars were taken following Lillegraven (1969: figure 5), and additions made for "condylarth" molars by Archibald (1982: figure 1). Multituberculate cusp terminology and measurements follow Clemens (1964) except for the /P4, where measurements were taken according to the system presented by Novacek and Clemens (1977).

Measurements of specimens are given in millimeters. Specimens were measured on an Ehrenreich Photo-Optical Industries "Shopscope" and rounded to the nearest one-hundredth of a millimeter.

# GEOLOGICAL AND PALYNOLOGICAL CORRELATIONS

The Hell Creek Formation of eastern Montana was deposited as the continental facies of the last major Late Cretaceous regression of the Western Interior Epeiric Sea (Frye, 1969; Cherven and Jacob, 1985). The overlying Tullock Formation is also continental in origin and was deposited during a transgression of the same sea (Cherven and Jacob, 1985) (see Figure 2). Both formations were deposited by fluvial systems on the proximal part of deltas formed from debris eroded primarily from the Elkhorn Mountains volcanic arc in southwestern Montana (Gill and Cobban, 1973; Cherven and Jacob, 1985).

Fluvial paleoenvironments during the K-T transition in eastern Montana were reconstructed by Fastovsky (1987). The results of this analysis indicate that:

> a broad fluvial plain, dissected by meandering, sediment-rich, suspended-load rivers, was aggraded by recycled, volcanic-rich clastic sediment. These were deposited by avulsive and sweeping channel migration, as well as by suspension settling during periodic floods. Vegetation was abundant, and collected in depressions, forming layers of matted plant debris. A high, fluctuating water table promoted the formation of weakly developed gleyed soils in moderately unstable flood-plain environments... A rise in water table, probably attributable to increased rainfall, induced a change in the paleoenvironments of this region at a time approximately coincident with that of the faunal and floral changes that define the K-P boundary. The elevated water table caused the gradual appearance of large ponded deposits and peat swamps, which resulted in discontinuous coal layers that characterize the K-P boundary.
> (Fastovsky, 1987:292)

The discontinuous coal layers produced by peat swamps are employed as a criterion for recognition of the lithostratigraphic boundary between the Hell Creek and Tullock formations (Calvert, 1912; Brown, 1952). In any single stratigraphic section, the formation boundary is defined as the first coal above the highest occurrence of dinosaurs (Brown, 1952). This "lowest" coal is termed the "Z," employing the letter terminology for designating coal beds in McCone County developed by Collier and Knechtel (1939). In this scheme, each stratigraphically higher coal bed receives a successive letter designation: Y, X, W, V, U. This system of lettering, especially the "Z" coal designation, has been employed in biostratigraphic studies in

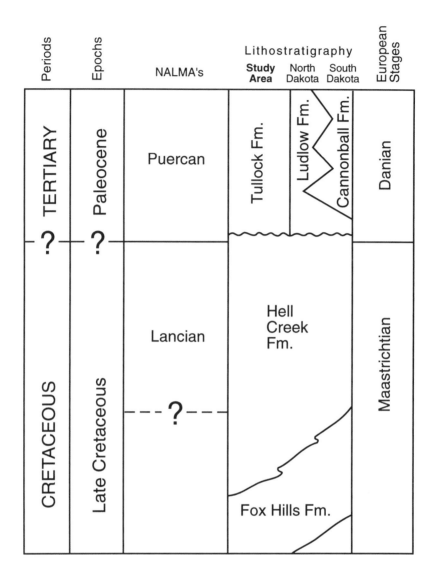

FIGURE 2. Correlation chart (modified from Bryant, 1989: figure 2), showing lithologic units in the study area (McCone County, Montana) and North and South Dakota. Correlation of lithologic units, geochronologic units, and NALMA's (North American Land Mammal "Ages" from Wood et al., 1941 and Russell, 1975) based on Archibald and Lofgren, 1990, and this study.

eastern Montana (Sloan and Van Valen, 1965; and many others). However, use of this terminology in more than a limited area without qualification can cause confusion because it implies regional continuity of each coal bed. When originally proposed, it was stated that this lettering system was not suitable for application in regional correlation: "Bed Z, is probably not a continuous bed but rather a succession of lenses of coal in about the same stratigraphic position" (Collier and Knechtel, 1939:18). And with respect to coals of the Tullock Formation, "correlation of coal beds from place to place is diffi-cult, and beds V, W, X, and Y, are all probably discontinuous, each being merely several lenses at about the same horizon" (ibid., p. 19). Therefore the employment of coal beds as regional correlation tools is problematic because coal beds in eastern Montana can be demonstrably discontinuous (Sholes and Cole, 1981; Archibald, 1982; Fastovsky, 1987).

At McGuire Creek the formational boundary is also recognized as the base of the first coal above the last dinosaurian fossils, but to eliminate potential confusion in correlation, this coal is mapped and designated the McGuire Creek Lignite (MCZ) (Plates 1-4), not the "Z" coal. Therefore, the MCZ is a local stratigraphic unit and has no implied regional significance.

In the McGuire Creek area, the upper Hell Creek Formation is composed primarily of gray and light to dark green sandstones, siltstones, and mudstones. Lesser amounts of purple or dark to light brown siltstones and mudstones are also present. Weakly lithi-fied, very thick (15-20 m) gray sandstones, containing Bug Creek assemblages, are local-ly abundant in the uppermost 30 m. These sandstones are overlain by a thick and later-ally persistent coal, the McGuire Creek Lignite (MCZ), which defines the base of the overlying Tullock Formation. Organic-rich sediments are rare in the upper Hell Creek Formation and usually consist of thin beds of black carbonaceous shale which have lim-ited lateral continuity. However, a thick shaly coal, the Tonstein Lignite (TL), is present about 30 m below the MCZ and is laterally traceable for over 4 km (Plate 1). Studies of the Hell Creek Formation in eastern Montana and the Dakotas are presented by B. Brown, 1907; R. Brown, 1952; Jensen and Varnes, 1964; Frye, 1969; Moore, 1976; Butler 1980; Archibald, 1982; Fastovsky, 1987; and Rigby and Rigby, 1990.

The Tullock Formation at McGuire Creek is composed primarily of gray, yellow, and light brown sandstones, siltstones, and mudstones. In contrast to the Hell Creek Formation, well developed coal beds are common in the Tullock Formation. Tullock sediments are, in general, more resistant to erosion, more tabular and more laterally continuous than Hell Creek beds. The difference in erosive properties is clearly evident in topographic expression in the McGuire Creek area; major cliff development is commonly associated only with the Tullock Formation (Plate 1). Additional descriptions of the Tullock Formation in eastern Montana are provided by Rogers and Lee, 1923; Collier and Knechtel, 1939; Archibald, 1982; Fastovsky, 1987; and Rigby and Rigby, 1990.


FACIES ANALYSIS AND GEOLOGICAL CORRELATIONS

Fastovsky (1987) has published a detailed facies analysis of vertebrate-bearing strata spanning the K-T transition in parts of eastern Montana. General classes of lithofacies

| This Study | Fastovsky and Dott, 1986 | Fastovsky, 1987 | Facies Interpretation |
|---|---|---|---|
| FACIES A | FACIES B | Siltstone Facies | Floodplain Deposits with Paleosols |
| FACIES B | ——— | ——— | Abandoned Channel Fill Deposits |
| FACIES C | FACIES D | Facies of Organic Accumulation | Coal-swamp Deposits |
| FACIES D | FACIES E | Variegated Facies | Pond Deposits |
| FACIES E | FACIES C | Facies of Epsilon Cross-Stratification | Laterally Accreted Channel Fill Deposits |
| FACIES F | FACIES A | Cross-Stratified Sandstone Facies | Vertically Accreted Channel Fill Deposits |
| FACIES G | ——— | ——— | Proximal (to Channel) Floodplain Deposits |

FIGURE 3. Comparison of terminology employed in this and other analyses of lithofacies in the upper Hell Creek and lower Tullock formations of eastern Montana.

at Bug Creek (Fastovsky and Dott, 1986; Fastovsky, 1987), which is only 1 km north of the McGuire Creek study area, can be recognized and employed at McGuire Creek (see Figure 3). Additions or modification of these facies based on strata at McGuire Creek are given below when necessary. Two additional lithofacies (Facies B and G) were recognized at McGuire Creek and are described in detail.

FACIES A
This facies is lithologically similar in part to the Siltstone Facies of Fastovsky, 1987, and Facies B of Fastovsky and Dott, 1986.

*Description*: Facies A is composed primarily of very thick to medium interbeds of green, gray, and purple, laminated to massive, siltstones and mudstones. Thick (up to 70 cm) lenticular sandstones that fine upward to siltstone or mudstone are also present but rare. Isolated and associated remains of larger-bodied vertebrates, primarily dinosaurs, are present. These occurrences usually are limited to isolated dinosaurian skeletal elements that are widely separated both geographically and stratigraphically. However, occurrences of more complete dinosaurian material were observed. The eroded and scattered remains of a partial ceratopsian skull are present in Section D (Plate 2). Mammalian remains were not recovered from this facies.

*Interpretation*: Facies A includes floodplain deposits that underwent soil development (Fastovsky, 1987; Fastovsky and McSweeney, 1987). Vertebrate fossils probably represent animal remains that were lying on the floodplain surface or in localized depressions that were subsequently buried by suspended sediment during flooding events. Fossils found in this depositional setting are unlikely to have been reworked from older strata, especially in the case of the ceratopsian skull cited above.

The absence of mammalian and other small-bodied vertebrate remains may be a reflection of the size of skeletal elements. Small bones have high surface-to-volume ratios and would be more readily destroyed during soil formation (Retallack, 1984).

## FACIES B

*Description*: Thin to medium interbeds of dark brown, brown, and gray sandstone, siltstone, and mudstone are present. Lenticular, fine-grained sandstones comprise 30-40% of this facies. Sandstone lenses are usually 15-20 cm thick and 15-20 m wide, and the largest observed is 1 m thick and 30 m wide. Sandstones also occur as thin laminated beds interbedded with siltstones and mudstones. Internally, sandstones exhibit ripple and small-scale tangential, planar, and trough (rare) cross-stratification sets. Siltstones and mudstones are laminated or massive, and all lithologies commonly have organic debris and plant macrofossils on laminae. Root casts are present but are rare. Siltstones and mudstones are more laterally continuous than sandstones, and usually can be traced laterally for over 25 m. Facies B fines upward and is overlain by Facies C.

The best exposures of Facies B near Section V (Plates 1, 2) indicate a channel-form morphology with steep bank margins which are deeply entrenched into Facies A beds. The base of Facies B is not exposed here, but a minimum of 6 m of relief is developed on this erosion surface, which dips 25-30 degrees to the east. The orientation of Facies B beds adjacent to this surface conforms to the eastward slope but levels out when traced toward the axis of the channel. This Facies B and Facies A contact trends 165 degrees and can be traced for 1.5 km to the southeast (Plate 1). Exposures of Facies B adjacent to Section V are over 150 m wide. East of these exposures are poorly exposed cross-stratified sandstones (Plate 1).

Articulated and associated remains of aquatic vertebrates including turtles, crocodiles, and champsosaurs are present in Facies B. Mammal and dinosaur remains were not recovered from this facies.

*Interpretation*: Facies B represents channels that were abandoned and slowly filled with suspended silt and mud. Lenses of cross-stratified sand indicate periodic influx of sand as channelized traction load during flooding events. The abundance of plant fossils but rarity of root traces suggests that the abandoned channel usually contained standing water. The presence of aquatic vertebrates supports this interpretation. Cross-stratified sandstones lateral (east) of Facies B outcrops (Section V) probably represent previously deposited thalweg channel-fill material.

## FACIES C

Facies C is lithologically similar to the Facies of Organic Accumulation of Fastovsky, 1987, and Facies D of Fastovsky and Dott, 1986.

*Description*: Facies C is composed of very thick to medium-bedded black to purple lignite and carbonaceous shale, containing 1-10 cm thick mud partings. Discontinuous or laterally persistent, 2-5 cm thick volcanic ash layers are present in both the McGuire Creek Lignite (MCZ) and Tonstein Lignite (TL). The MCZ and TL each contain a particularly persistent volcanic ash layer whose presence was useful for mapping across disjunct exposures. The MCZ is locally overlain by a 20-50 cm thick, massive, black to dark gray mudstone which is devoid of plant macrofossils and roots. This mudstone is useful for correlation of the MCZ between isolated exposures. Amber granules are abundant in the TL. Vertebrate remains were not recovered from Facies C.

*Interpretation*: Facies C represents the development of widespread coal-swamp deposits in the McGuire Creek area. Extensive coal-swamp development is evident because the MCZ and TL are laterally traceable over more than 20 and 4 sq. km, respectively (Plate 1). The extensive development of lignites in the McGuire Creek study area is noteworthy because coal horizons at or below the formation boundary can be extremely localized elsewhere (Archibald, 1982; Fastovsky and Dott, 1986; Fastovsky, 1987).

It appears that the TL, and possibly the MCZ, may have significant utility as correlation tools in biostratigraphic studies because they are evidently present in regions north and west of the McGuire Creek study area. A Hell Creek Formation lignite, the Null Coal, mapped within the Bug Creek drainage by Smit et al. (1987) and Rigby and Rigby (1990), almost certainly is equivalent to the TL based on elevation, stratigraphic position, and general lithologic similarity. However, the absence of ash layers in the Null Coal precludes definite correlation (NW quadrant of Section 16, Plate 1, and directly below the Bug Creek Anthills channel fill, NW quadrant of Section 9, T 22 N, R 43 E).

A coal in the upper Hell Creek Formation containing one or two ash layers and amber is exposed at Jawbone Coulee and Thomas Ranch, 5 and 15 km west of the McGuire Creek study area, respectively. The Jawbone Coulee lignite has two ash layers and amber, and closely parallels the TL in general stratigraphic position, lithologic character, and position of ashes within the coal (Section 13, T 22 N, R 42 E). In the opinion of Dr. Carol Hotton (pers. comm., 1989), the TL may also be equivalent to the Tonstein Coal at the Thomas Ranch locality (Hotton, 1988; Section 14, T 22 N, R 41 E) based on approximate stratigraphic position, presence of an ash layer, and abundant amber granules.

The MCZ is difficult to trace into the northern part of the study area, and correlations become questionable (Plate 1). Correlation between lignites is especially difficult to the northwest (Section HH) because of discontinuous exposures and thinning of coals. It appears that the MCZ and upper "Z" coal of the lower Bug Creek drainage (not including Russell Basin) are two different units (Section OO, Plate 4). However, mapping north of Section DD (Plate 4) indicates that the MCZ is traceable northward into the headwaters of the Bug Creek drainage and is equivalent to the uppermost lignite of the "Z" complex at Russell Basin (upper "Z" of Smit et al., 1987 and Rigby and Rigby, 1990). It appears that the upper "Z" and MCZ (Section OO) are two units of a series of thin lignites that comprise the upper "Z" complex of Russell Basin.

## FACIES D

Facies D is lithologically similar to the Variegated Facies of Fastovsky, 1987; Facies E of Fastovsky and Dott, 1986; and the "variegated" beds of Archibald, 1982.

*Description*: Facies D is composed of thin to medium interbeds of yellow-brown or gray laminated siltstone and mudstone. Thin individual beds can be traced for hundreds of meters laterally in continuous exposures. Facies D, as a whole, is laterally traceable for over 25 sq. km in the McGuire Creek area. Facies D contains isolated occurrences of articulated elements of turtles and champsosaurs.

*Interpretation*: Facies D represents extensive pond deposits (Fastovsky, 1987).

## FACIES E

Facies E is lithologically similar to the Facies of Epsilon Cross-Stratification of Fastovsky, 1987, and Facies C of Fastovsky and Dott, 1986.

*Description*: Facies E is composed of 3-20 m thick gray sandstones in elongate sheet morphologies whose base is scoured into Facies A or another Facies E unit (Plate 3). Basal scours are commonly flat but can also exhibit over 5 m of relief (Little Roundtop Channel, Sections Q and 861, Plate 3). Weakly to well-defined inclined heterolithic stratification (IHS) (following Thomas et al., 1987, this term replaces "epsilon cross-stratification") sets are present that exceed 15 m in thickness and exhibit dips of 10-20 degrees. IHS is defined by alternating beds of sandstone, siltstone, and mudstone. It extends down to the basal scour in some cases (Little Roundtop Channel, Plate 3), but in others (Second Level Channel, Plates 3 and 4) the basal 1-7 m is composed of thick, vertically stacked sets of trough cross-stratified sandstone (similar to Facies F) which fine upward in both grain size and scale of sedimentary structures before interfingering with IHS (Sections 861, Q, R, Plate 3; Section T, Plate 2; Sections R, 871, Plate 4; Sections O, P, Plate 4). Paleocurrents of trough cross-strata are approximately perpendicular to IHS surfaces (Figure 4). Complete Facies E sequences fine up to siltstone and mudstone and are overlain by Facies C, or Facies G and C.

Poorly sorted, clast- to matrix-supported lag concentrations of clayball conglomerate, carbonized plant debris, fossil wood, and vertebrate fossils are commonly present at the base of Facies E channels. Basal lags consist of discontinuous 30-100 cm thick lenses of poorly sorted, clast-supported, pebble-cobble conglomerates composed of rounded mudstone and siltstone clasts. Conglomerates also occur as lenses within or underlying trough cross-bedded, medium to fine grained sandstone. Fragments and impressions of carbonized plant debris are abundant, while fossilized wood is rare. Localized depressions in channel floors are commonly filled with conglomerates containing rich concentrations of microvertebrate fossils. These fossil remains usually consist of isolated teeth and disarticulated elements (complete elements are rare) of aquatic and terrestrial vertebrates which have undergone hydraulic transport and sorting. Fossils are commonly fragmentary and abraded, but many pristine bones and teeth were also recovered. Fossils are intermixed with pebble-sized intraclasts that apparently have similar settling velocities (this is not true for larger fossils). Isolated teeth of mammals and dinosaurs are common, and fragments of shed ceratopsian teeth com-

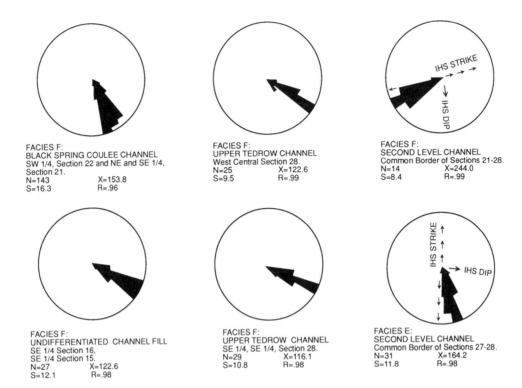

FIGURE 4. Trough cross-bedding orientations measured in the upper Hell Creek Formation at McGuire Creek. Data presented are grouped by mapped channel fill except where noted. N= number of measurements; X= vector mean in degrees; S= circular dispersion in degrees; R= vector magnitude. See Plate 1 for location of channel fills measured.

prise the great majority of dinosaurian dental remains. Mammal jaws are uncommon, and mammalian maxillary fragments are rare. In comparison to the number of dental remains at any one site, mammalian postcranial elements are under represented and usually consist of dense elements such as phalanges, carpals, tarsals, and distal or proximal fragments of other limb bones. Vertebrate fossils were not found on IHS surfaces or immediately above the base of Facies E sequences.

Facies E lag concentrations which produce Bug Creek vertebrate assemblages are the Brown-Grey, Second Level, and Little Roundtop channel fills. The Z-Line Channel does not cut deeply into Facies A beds and yields a Puercan assemblage. Well-preserved teeth and many heavily abraded fragments of dinosaur bone are present at locality V87072 (Brown-Grey Channel).

*Interpretation:* Facies E represents laterally accreted channel-fill deposits. IHS is interpreted as lateral accretion surfaces of point-bar deposits (Fastovsky, 1987), while thick trough cross-strata in the lower region of Facies E channel fills represent vertically accreted thalweg sandstones (Mossup and Flach, 1983). Facies E channels were entrenched into pre-existing Facies A units. Superposed Facies E fills (Little Roundtop and Second Level channels, Plates 3 and 4) indicate that multiple channel cutting and filling events also occurred. Therefore, considerable reworking of pre-existing strata and their entombed vertebrate remains probably occurred.

## FACIES F

Facies F is lithologically similar to the Cross-Stratified Sandstone Facies of Fastovsky, 1987, and Facies A of Fastovsky and Dott, 1986.

*Description*: Facies F is composed of 10-20 m thick, multi-storied, gray sand bodies which are commonly overlain by Facies G and C, and usually sharply bounded below by Facies A. Facies F units truncate Facies E channels and also Facies D and C deposits above the MCZ (Plates 2-4). Relief on these erosion surfaces exceeds 15 m (Sections II and 811 or Sections O, P, 856, and 855, Plate 4; Sections H-B, Plate 2).

Individual stories are lenticular, have a sharp undulatory base, and are about 2-10 m thick and 10-50 m wide. Successive stories are commonly incised into preceding channel-fill cycles, and several erosion events can be seen in good exposures (for example, Section H, Plate 2). Limited three-dimensional exposure of some stories shows that they have channel-form geometries with subparallel axes. Paleocurrent data indicates a south-southeast flow direction with a markedly unidirectional trend (Figure 4). Mapped Facies F sand bodies exhibit widths of approximately 0.5 km perpendicular to paleocurrent means (Plate 1).

Relatively complete stories exhibit a decrease in scale of sedimentary structures from very thick or thick sets of trough and planar cross-stratification to medium or thin sets of trough, planar, and ripple cross-stratification, overlain by siltstone. Sediments within most stories fine upward from medium grained sandstone with minor amounts of mudstone pebble conglomerate at the base to organic rich fine-grained sandstone or siltstone at the top.

Basal scours of Facies F sand bodies commonly exhibit localized lag concentrations of poorly sorted, clast- and matrix-supported, clayball conglomerates which contain fossil wood, carbonized plant debris, and vertebrate fossils (see Facies E for a general description of the depositional setting of these vertebrate remains). In Facies F, vertebrate fossil remains were rarely found above the basal lag of the lowest story.

Facies F lag concentrations which produce Bug Creek assemblages are the Black Spring Coulee and possibly the Upper Tedrow and "Lower Tedrow" channel fills. Jacks Channel produces a Puercan vertebrate assemblage. Vertebrate fossil occurrences in lag concentrations usually consist of isolated elements, but abundant well-preserved dinosaur material is present at Black Spring Coulee (V87030; see the following chapter on Reworking of Fossils for discussion).

*Interpretation*: Facies F multi-storied sand bodies were probably developed through vertical aggradation in laterally stable channels of low sinuosity. The subparallel channel-form geometries of each story, low divergence of paleocurrents, and absence of lateral accretion features such as IHS support this interpretation. Upward decrease in grain size and scale of sedimentary structures within each story records waning paleoflow strength between flood events. Multiple scour and fill events occurred within channels, as evidenced by entrenchment of superposed stories. The deep entrenchment of Facies F channels into pre-existing sediments suggests that considerable reworking of fossils occurred.

Other channel fills, not assignable to a channel facies because of poor exposures, contain important occurrences of vertebrate fossils. The K-Mark Channel yields a Lancian vertebrate assemblage. Matt's Dino Quarry Channel produces disarticulated but closely associated hadrosaur skeletal elements and a complete turtle shell. The Shiprock and Up-Up-the-Creek sites yield Bug Creek assemblages.

FACIES G

*Description*: Facies G is composed of tabular interbeds of thin to thick brown, gray, and greenish-gray sandstone, siltstone, and mudstone. Sandstones comprise 20-30% of this facies and are commonly laminated, but also exhibit ripple and small scale planar cross-stratification. Sandstones also occur as elongate lenses up to 10 cm thick within siltstones. Siltstones and mudstones are laminated and contain organic debris and well-preserved plant macrofossils on laminae. Complete leaves are extremely abundant locally in all laminated lithologies, and large palm fronds were observed. Fossilized in situ tree stumps are present at sections B and H (Plate 2). Root casts are present, but are not common. Individual sandstone, siltstone, and mudstone units are commonly traceable over tens of meters.

Thin black carbonaceous shales-mudstones composed of fissile mats of laminated plant debris are also present, and are traceable for hundreds of meters in continuous exposures. Articulated and associated remains of turtles, champsosaurs, and crocodiles are abundant in carbonaceous shales-mudstones. Neither mammals nor dinosaurs were recovered from this facies.

Facies G overlies floodplain facies (Facies A, east part of Plate 2) and overlies and interfingers with channel facies (Facies E and F, Plates 2-4). Facies G can be a continuum with Facies A, E, and F and boundary recognition must be arbitrarily determined in many cases. However, with respect to Facies G and A, they differ in color, and Facies G beds are more tabular and laterally continuous. Also, Facies G sediments contain well-preserved plant macrofossils (including in situ tree stumps) and internal bedding features (laminations and ripples).

*Interpretation*: Facies G represents floodplain deposits developed in association with channel facies. These floodplain deposits comprise the culmination of fining upward sequences of channel facies (E and F), or the lateral floodbasin deposits of these channels. Extensive but short-lived swamps were developed on floodplains, as evidenced by the thin but laterally continuous carbonaceous shales containing abundant aquatic vertebrates. Cross-stratified sandstones probably represent minor crevasse-splay development on the floodplains. The presence of tree stumps, abundant plant macrofossils, and root traces indicates that these floodplains were probably heavily vegetated. However, paleosol development was generally not as advanced as in Facies A because delicate sedimentary structures such as laminations and whole-leaf plant fossils are extremely abundant and well preserved.

## GEOLOGIC HISTORY OF THE MCGUIRE CREEK AREA

A geologic history spanning the upper Hell Creek and lower Tullock formations at the McGuire Creek area can be constructed based on superposition, cross-cutting relationships, and interfingering between the facies described above. This sequence of events applies only to the study area at McGuire Creek.

Sediments within the upper but not uppermost Hell Creek Formation at McGuire Creek consist primarily of floodplain facies (Facies A; Plates 2-4). Floodplain facies also contain evidence of small-scale stream and extensive peat swamp development (e.g. TL: Tonstein Lignite). In situ dinosaur remains above and below the TL indicate that dinosaurs inhabited these floodplains.

The predominantly floodplain deposits of Facies A were then scoured by a series of major streams. Channels created by these streams were filled by sediments representing Facies B, E, and F. The thickness of IHS sets (up to 10 m; Sections 861, Q, R, Plate 3) and of the channel fills themselves (up to 20 m; Section II, Plate 4) indicate that these streams were very large. Over 15 m of relief is present on the contact between floodplains (Facies A) and channel-fill facies (Facies B, E, F; see Sections II through 811, Plate 4; and Sections H through B, Plate 2). The 15 m of relief indicates that stream channels were deeply entrenched into floodplains. As a result, the erosion surface developed on floodplain deposits (Facies A) is traceable for over 3 km (Sections K through R, Plate 3; Sections R through II and O through 855, Plate 4). An immense volume of previously deposited strata (mostly Facies A) was removed during these channeling events, and the likelihood that vertebrate remains were reworked is high. Channels were filled by vertically accreted (Facies F) or laterally accreted (Facies E) sandy sediment, while others were abandoned and filled primarily by finer-grained sediment that settled from suspension (Facies B).

A complex record of superpositional and cross-cutting relationships between laterally accreted (Facies E) and vertically accreted (Facies F) channels indicates the occurrence of multiple channel and fill events within this stratigraphic interval (below MCZ, above TL). Channels were deeply entrenched into pre-existing channel facies as well as floodplains (Facies A), and considerable reworking of vertebrate remains must have occurred. Eight mapped channel-fill units, six of which represent individual channel fills, are evidence of this complexity (Plates 1-4). Excellent exposures permit temporal sequencing of a limited number of these channeling events (a simplified version of channel-fill relationships is presented in Figure 5 and should be referred to in the following discussion). Superpositional and cross-cutting relationships indicate that the oldest demonstrable channel fill (of the six that were individually mapped) is the Little Roundtop Channel, which is scoured and overlain by the Second Level Channel (Plates 3, 4). The Second Level Channel is cut by both the Black Spring Coulee and Upper Tedrow Channel fills (Plates 3, 4). The Upper Tedrow channel cuts both the Brown-Grey and "Lower Tedrow" channel fills (Plate 3) ("Lower Tedrow" and Upper Tedrow channels may be equivalent, see section below on The Tedrow Area). These cross-cutting relationships reveal that both the Upper Tedrow and Black Spring Coulee channel fills are younger than the Little

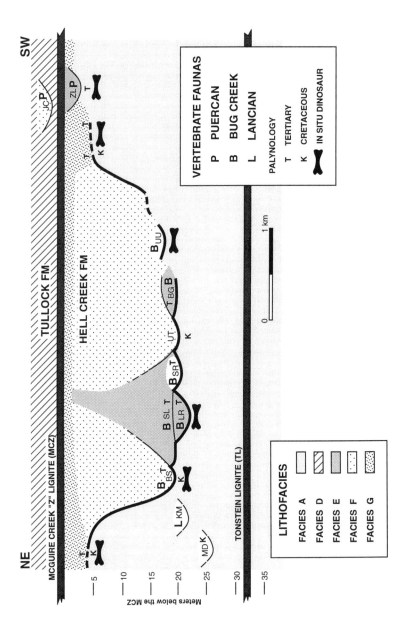

FIGURE 5. Schematic stratigraphic relationships of lignites, channel fills yielding vertebrate faunas, floodplain deposits with in situ dinosaur remains (horizontal long-bone figures), sediment samples analyzed for pollen, and age interpretation of palynofloras. Horizontal distance depicted is approximately 5 km in NE-SW orientation. Channel fills (or Local faunas) are: (MD) Matt's Dino Quarry; (KM) K-Mark; (BS) Black Spring Coulee; (LR) Little Roundtop; (SL) Second Level; (SR) Shiprock; (UT) Upper Tedrow; (BG) Brown-Grey Channel; (UU) Up-Up-the-Creek; (ZL) Z-Line; (JC) Jacks Channel. KM and MD are not superpositionally overlain by the MCZ as shown, but are actually at the tops of isolated buttes. KM, MD, UU, and SR could be either Facies E or F. Heavy line denotes the erosion surface created by successive channeling events.

Roundtop Channel. Temporal relationships that remain uncertain are those between the Black Spring Coulee and Upper Tedrow, Brown-Grey and Little Roundtop, and Brown-Grey and Second Level channel fills. Physical relationships of the Z-Line Channel to any of the other five channel fills cannot be determined.

Three types of vertebrate assemblages (Lancian, Bug Creek, and Puercan) were collected from lag deposits in Facies E, Facies F, or undifferentiated channel fills within this stratigraphic interval. Z-Line (Facies E) is the only channel fill that preserves a Puercan fauna below the MCZ (refer to Figure 5). Bug Creek assemblages are present in the Little Roundtop, Second Level, Black Spring Coulee, and Brown-Grey channel fills, and the Shiprock and Up-Up-the-Creek sites. A Lancian assemblage is present only in the K-Mark Channel, an erosional remnant of a channel fill of unknown size and facies (Plate 3). Matt's Dino Quarry yielded an associated hadrosaur skeleton in a channel fill of undetermined facies (Plate 3), but mammals were not recovered.

The presence of Lancian, Bug Creek, and Puercan vertebrate assemblages in nearly identical facies in the same stratigraphic interval undoubtedly indicates that some degree of faunal change was occurring at this time. However, because of discontinuous exposures and the lenticular nature of channel facies, temporal sequencing of the three assemblages by superpositional or cross-cutting relationships cannot be determined. Interpretations of faunal change at this stratigraphic interval are ambiguous because they cannot be clearly supported by a superposed stratigraphic sequence of faunas.

Channel fills (Facies B, E, F) commonly fine upward and/or grade laterally into floodplain deposits (Facies G). These floodplains were accreted in unison with channel migration (Facies E) or were developed after vertically accreted channels (Facies B, F) were filled. Facies G floodplains also overlie Facies A floodplains. Areas that preserve this relationship are Sections II and 811 (Plate 4; Figure 5, base of BS traced to left) and Sections B through H (Plate 3; Figure 5, just right of UU, then traced to right). These sections show that channels (Facies F) which produce Bug Creek assemblages (Black Spring Coulee Channel) interfinger with floodplain deposits (Facies G) that contain aquatic vertebrates but lack dinosaurs. In the absence of a facies bias, it appears that the lack of dinosaurs in Facies G floodplains indicates that they were extinct by this time, at least locally. Paradoxically, where it can be determined, floodplains (Facies G) that lack dinosaur remains appear to have accreted in unison with channels (Facies F) that yield dinosaurs in lag concentrations (Section II, Black Spring Coulee Channel, Plate 4; Sections H-B, Plate 2). However, the presence of dinosaur remains in these channel fills is probably the result of reworking (see Lofgren at al., 1990). Reworking would explain why disarticulated remains of dinosaurs are present in channel facies (Facies F) but are absent in contemporaneous floodplain facies (Facies G).

Based on topographic elevation, the stratigraphic level of the basal lags of channels (Facies E, F) which contain Bug Creek assemblages are equivalent to floodplains (Facies A) containing in situ dinosaurs. However, lateral correlation of Facies F channels containing Bug Creek assemblages indicates that these channels interfinger with Facies G floodplains, not Facies A floodplains (Sections II through 8111, Plate 4; Sections B through H, Plate 2). This is a reflection of the deep entrenchment of channels into older floodplain deposits (Facies A), where as much as 20 m of relief is observed. Therefore, correlation of facies provides a more reliable method of determining relative age than elevation or relative (to laterally persistent coals) stratigraphic level. Analysis of facies relationships of vertebrate-bearing fluvial deposits is critical because faunas present at the same stratigraphic level or elevation could differ significantly in age.

In contrast to the examples where channels containing Bug Creek assemblages can be traced into age-equivalent floodplains, the relationship of channel fills bearing Puercan or Lancian vertebrate assemblages to age-equivalent floodplain deposits (Facies A or G) cannot be determined. The K-Mark Channel and Matt's Dino Quarry site are located at the top of isolated buttes, making lateral correlation to floodplain deposits impossible. The margins of the Z-Line Channel (Facies E) are not exposed in contact with floodplain deposits.

Complete channel-fill (Facies B, E, F) sequences fine upward into floodplains (Facies G) and are all overlain by a laterally extensive lignite (Facies C), the MCZ (base of the Tullock Formation). This indicates that, following the episode of major channeling and erosion, a large peat swamp was developed which blanketed most of the study area. This swamp was in turn replaced by development of extensive areas of standing water, represented by thick sequences of pond deposits (Facies D).

Pond deposits (Facies D) are overlain by fluvial sediments that aggraded in various depositional settings. Of interest are channels (Facies F) with lag concentrations containing vertebrate fossils that are deeply entrenched (up to 15 m) into pond deposits (Facies D) overlying the MCZ (Facies C). These channels, for example, Jacks Channel (Plates 2, 3), contain Puercan assemblages (which lack dinosaurs and typical Lancian mammals). Remains of dinosaurs are not present above the MCZ in the McGuire Creek area, and apparently dinosaurs were extinct by this time (but see Rigby, 1989).

## PALYNOLOGICAL CORRELATION OF MCGUIRE CREEK LITHOFACIES

It has long been known that a palynological change that could be recognized in stratigraphic sequences in eastern Montana was roughly coincident with R. Brown's (1952) formula (first coal above highest dinosaur) for the terrestrial K-T boundary (Norton and Hall, 1969; Oltz, 1969; Tschudy, 1970). Based on more recent refinements, Cretaceous strata can be differentiated from Tertiary strata by the absence or great reduction in abundance of Cretaceous indicator species in the latter (Smit et al., 1987; Hotton, 1988). Therefore, palynology can be used for determining the relative age of vertebrate-bearing strata in the upper Hell Creek Formation where lithostratigraphic relationships are unclear, if one facies yields a Cretaceous palynoflora while another contains a Tertiary pollen assemblage. (For discussion of palynological recognition of the K-T boundary, see chapter below on Cretaceous-Tertiary Correlations). Thus, palynology is useful in correlating between disjunct exposures of channel (E, F, B) facies and/or floodplain facies (A, G). The results of palynological analyses of rock samples from McGuire Creek are given in Table 2.

Palynological correlation indicates that channel facies with Bug Creek or Puercan assemblages (Facies E, F) are younger than floodplain facies containing in situ dinosaur

Table 2. Results of palynological analysis of selected rock samples from McGuire Creek

| | | LOCATION | | | |
|---|---|---|---|---|---|
| Rock Sample Field No. | Age[a] | Measured Section | Plate Number[b] | Facies | Channel Fill |
| 86DLL7-29-1 | K | FF | 3 | A? | — |
| 86DLL7-29-2 | K[c] | FF | 3 | A | — |
| 88DLL7-14-30 | P | GG | 3 | E | BG |
| 88DLL7-14-13 | P | GG | 3 | E | BG |
| 87DLL7-26-7 | P | M1 | 3 | E or F | SR |
| 87DLL7-2-1 | P | B | 2 | A | — |
| 87DLL7-2-3 | P | B | 2 | A | — |
| 87DLL7-16-21 | K | E | 2 | A | — |
| 87DLL7-17-2 | P | G | 2 | F | Undiff. |
| 87DLL8-3-2 | K | O | 4 | A | — |
| 87DLL8-3-3 | P | O | 4 | E | SL |
| 88DLL6-26-3 | K | AA | 3 | — | MD |
| 88DLL8-10-1 | P | 861 | 3 | E | SL |
| 88DLL7-15-6 | P | R | 3,4 | E | LR |
| 88DLL7-15-7 | P | R | 3,4 | E | LR |
| 88DLL7-28-2 | P | V | 2 | B | Undiff. |
| 87DLL7-28-8 | P | N | 2 | A | — |
| 87DLL7-28-4 | P | N | 2 | A | — |
| 88DLL7-18-4 | P | II[d] | 4 | F | BSC |
| 88DLL8-11-3 | K | II | 4 | A | — |
| 88DLL8-11-4 | K | II | 4 | F | BSC |
| 88DLL8-11-7 | K | II | 4 | F | BSC |

a. All rock samples except 86DLL7-29-2 were analyzed for pollen and spore content by Carol Hotton.

b. For stratigraphic position of rock samples within measured sections refer to the designated plate.

c. Prepared slides of 86DLL7-29-2 contained abundant grains of *Aquilapollenites* (analysis by the author) which indicate a Cretaceous age for the rock sample (Hotton, 1988).

d. For detailed taxonomic lists of the palynoflora identified in the rock samples from Section II, see Lofgren et al. (1990). Taxonomic listings of the palynofloras identified in the other rock samples will appear elsewhere.

Table 3.  Geochronologic age of local faunas or channel fills in the uppermost Hell Creek Formation at McGuire Creek, based on palynological analyses of rock samples

| Local Fauna or Channel Fill[a] | Vertebrate Assemblage[b] | Palynological Age[c] |
|---|---|---|
| Black Spring Coulee | BC | Paleocene * |
| Second Level | BC | Paleocene * |
| Little Roundtop | BC | Paleocene * |
| Shiprock | BC | Paleocene * |
| Brown-Grey | BC | Paleocene * |
| Upper Tedrow | BC? | Paleocene |
| K-Mark | L | Unknown |
| "Lower Tedrow" | BC? | Paleocene? or Cretaceous? |
| Z-Line | P | Paleocene |
| Up-Up-the-Creek | BC | Unknown |
| Matt's Dino Quarry (mammals absent) | L? | Cretaceous * |

a. The Shiprock and Up-Up-the-Creek local faunas are within undifferentiated channel fills.
b. Vertebrate assemblages present are Bug Creek (BC), Lancian (L), and Puercan (P).
c. Analysis of samples collected within channel fills indicated by *. Other channel fills were assigned ages based on superpositional or crosscutting relationships with beds of known palynological age.

remains (Facies A). Facies G floodplain deposits were not analyzed for pollen, but they are Paleocene, because they unequivocally overlie channel fills or Facies A floodplain deposits that produce Paleocene pollen. In contrast, Facies A floodplain deposits may be Paleocene or Cretaceous in age, depending upon the stratigraphic interval sampled. Where channel facies are deeply entrenched into Facies A floodplain deposits, these floodplain deposits are Cretaceous in age, based on pollen (Section FF, Plate 3, and Sections II and O, Plate 4). Where Facies G overlie Facies A floodplain deposits, the uppermost portion of Facies A deposits are palynologically Paleocene (Sections D and B, Plate 2). Smaller-scale channel entrenchment (less than 3 m) into Facies A floodplains containing Paleocene pollen occurs at Section N (Plate 2). Where sampled, Facies A beds that contain in situ dinosaurs produce Cretaceous pollen (Section O, Plate 4; the same bed in Section E from which rock sample 87DLL7-16-21 was collected produced a disarticulated ceratopsian skull 150 m to the south). In situ dinosaurs have not been found in Facies G floodplain deposits that yield Paleocene pollen (but these floodplain deposits yield turtles, champsosaurs, and crocodiles; Sections B-D, Plate 2). Without exception at McGuire Creek, in situ dinosaur remains have not been found in the stratigraphically lowest Paleocene floodplain deposits (as identified palynologically). This is also true in other K-T transition stratigraphic sequences of floodplain deposits in eastern Montana, where the highest in situ dinosaur remains are never found within 2 m (below) of the palynological boundary (Archibald, 1987c).

Channel fills containing vertebrate fossils in the uppermost Hell Creek Formation were sampled for pollen when suitable lithologies were present. Otherwise, ages of channel fills were determined by superpositional or cross-cutting relationships with another rock unit whose age was determined palynologically. The results of these analyses are given in Table 3 and Figure 5.

All channels containing Bug Creek or Puercan vertebrate assemblages are demonstrated to be of Paleocene age where palynological determination is possible (Table 3; Plates 2-4, Figure 5). Therefore, channel fills containing Bug Creek or Puercan assemblages are palynologically equivalent to floodplain deposits that lack in situ dinosaurs (Facies A and G) and are palynologically dissimilar to floodplain deposits that contain in situ dinosaurs (Facies A).

Similar conclusions, based on lithostratigraphy and palynology, are reached concerning the timing of geologic events in the uppermost Hell Creek Formation at McGuire Creek. Lithofacies correlation indicates that channels (Facies B, E, and F) are entrenched into floodplain deposits (Facies A) containing in situ dinosaur remains (Figure 5; Plates 2-4). Limited lateral correlation of channels (Facies F) suggests interfingering with floodplain deposits that lack dinosaurs (Facies G) (Figure 5; Plates 2, 4). It appears, on the basis of palynology, that deep entrenchment of Paleocene channels into a pre-existing landscape composed of mostly Cretaceous and less commonly Paleocene floodplain deposits occurred; this erosion surface denotes the palynological K-T boundary (shown by heavy line in Figure 5). Correlation of channel facies (Facies E, F, B) laterally to floodplain facies (Facies G) is corroborated by palynological analysis. If present, the uppermost parts of Facies A floodplain deposits that lack in situ dinosaur remains are Paleocene. These strata may be remnants of floodplain deposits accreted lateral to Paleocene channels whose isolated vestiges (for example, the Little Roundtop or Brown-Grey channels) cannot be traced laterally into floodplain sequences because of repeated channeling.

In no case at McGuire Creek is a Bug Creek or Puercan vertebrate assemblage clearly associated with a Cretaceous palynoflora. The basal lags of these channels may contain reworked mud pebbles, cobbles, and boulders that produce Cretaceous pollen (Lofgren et al., 1990) but the channel fills yield Paleocene pollen where sampled (Table 3; Figure 5).

Relative age discrimination of individual channel fills containing Lancian, Bug Creek, or Puercan assemblages in the upper Hell Creek Formation at McGuire Creek is severely limited because all but one yield Paleocene palynofloras (Table 3; Figure 5). Matt's Dino Quarry Channel contains associated hadrosaur skeletal remains (but lacks mammals), yields a Cretaceous palynoflora, and is the only site that is demonstrably older than the others listed in Table 3. The only identifiable Lancian site, the K-Mark Channel, probably is palynologically Cretaceous, but this cannot be confirmed because the sediments are unsuitable for pollen analysis.

## MCGUIRE CREEK BIOSTRATIGRAPHY

Development of biostratigraphic units at McGuire Creek that span the uppermost Hell Creek Formation is limited because of the lack of stratigraphic control on channel fills that yield fossil assemblages. References to Plates 2-4 are given to demonstrate biostratigraphic relationships in the following discussion (simplified version in Figure 5).

In spite of poor stratigraphic control, superpositional or cross-cutting relationships between channel facies or lignite beds (TL or MCZ) can be determined in some cases. The Jacks Channel local fauna (Puercan) is younger than the others listed in Table 3 because Jacks Channel lies entirely above the MCZ (Section EE, Plates 2 and 3). The other local faunas were collected from channel fills that are (or apparently were) capped by the MCZ prior to Holocene erosion.

Below the MCZ, limited temporal ordering of local faunas containing Bug Creek assemblages can be determined. The Little Roundtop Local Fauna is older than the Second Level Local Fauna because these respective channel fills are in superposition (Sections 861-R, Plate 3). The crosscutting relationship between the channel fills reveals that the Black Spring Coulee Local Fauna is younger than the Second Level Local Fauna (Section 871-II, Plate 4). The Upper Tedrow Channel cuts and/or overlies both the Second Level (Sections O-P-856-855, Plate 4) and Brown-Grey (Section GG, Plate 3) channels. The biostratigraphic utility of these relationships is limited, however, because only one mammal, *Meniscoessus robustus*, has been found in the Upper Tedrow Channel. The "Three Buttes Local Fauna" is an informal grouping of vertebrate assemblages found in the erosional remnants of channel fill(s) in an area adjacent to exposures mapped as Upper Tedrow or Second Level channels (Plate 1; NE quadrant, SE quadrant, Section 28, T 22N, R 43 E). However, lag deposits which produce fossils included within this "local fauna" are poorly constrained stratigraphically and cannot be correlated with either channel fill because of Holocene erosion of intervening exposures ("Three Buttes Local Fauna" is placed in quotations in Appendix 2 to denote this uncertainty).

The remaining channel fills containing the Lancian (K-Mark, Section BB, Plate 3), other Bug Creek (Shiprock, Section M1, Plate 3; Up-Up-the-Creek, Plate 1), and Puercan (Z-Line, Sections JJ-N-W, Plate 2) local faunas lack stratigraphic control, making their temporal relationships uncertain (other than relative to Jacks Channel). Usually lateral correlation of strata that cap each channel fill would aid in determining relative age, but all upper Hell Creek channel fills are (or apparently were) capped by the same laterally traceable stratigraphic unit, the MCZ.

Palynological data adds little information useful for more detailed biostratigraphic zonation of the uppermost Hell Creek Formation because all sites yielding vertebrates (with the exception of Matt's Dino Quarry Channel) yield similar palynofloras (Table 3, Figure 5).

In summary, it is virtually impossible to develop a local biostratigraphic zonation for the upper Hell Creek Formation at McGuire Creek that orders the three kinds of vertebrate assemblages. The uppermost 25 m of the Hell Creek Formation contains the records of a major faunal turnover, but the transition is recorded within channel-filling events that are not amenable to biostratigraphic subdivision.

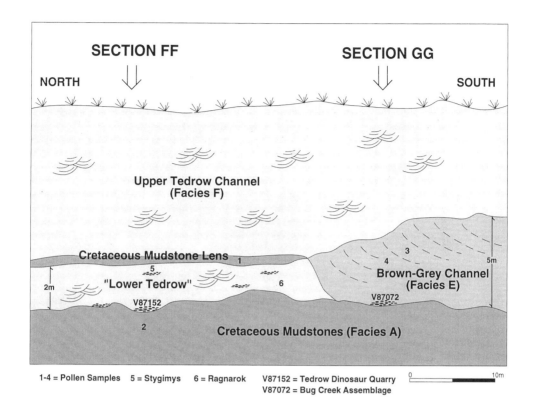

FIGURE 6. Idealized cross-section of the Tedrow Area showing location of fossil and rock sample sites, channel fills, and contacts between depositional units. The position of measured sections FF and GG are shown on Plate 3. Rock samples analyzed for pollen are: (1) 86DLL7-29-1: Cretaceous; (2) 86DLL7-29-2: Cretaceous; (3) 88DLL7-14-30: Paleocene; (4) 88DLL7-14-13: Paleocene. Isolated mammal specimens are: (5) *Stygimys* incisors. (6) *Ragnarok* lower molars.

## THE TEDROW AREA

The complexity of channel cutting and filling in the uppermost Hell Creek Formation is shown in the Tedrow area (the term "Tedrow area" describes exposures adjacent to and including Sections FF-GG, Plate 3; Figure 6). The Brown-Grey, Upper Tedrow, and "Lower Tedrow" channel fills, and a mudstone lens containing Cretaceous pollen, are exposed in the Tedrow area, but determination of a sequence of depositional events for the channel fills and mudstone lens is difficult because Holocene erosion has removed critical information. These difficulties highlight the uncertainty inherent in biostratigraphic study of complex fluvial channel deposits, even in an area in which three-dimensional exposures are available.

The "Lower Tedrow" Channel is a trough cross-stratified sandstone containing a sparsely sampled vertebrate fauna consisting of associated ceratopsian skeletal remains (Figure 7) and a few isolated mammal teeth referable to *Ragnarok* and

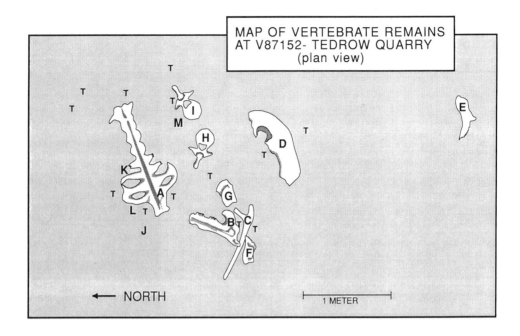

FIGURE 7. Plan view of Tedrow Dinosaur Quarry (V87152). Letters on quarry map identify the following elements: (A) Sacrum: broken at base of neural arch over nearly entire length, entire dorsal part of sacrum absent. Transverse processes missing on many vertebrae. (B) Mandible (right): edentulous; dentary, surangular, and possibly angular present. Dental battery fractured-weathered. (C) Dorsal rib #2 (left): distal end of shaft lost. (D) Squamosal (right): lateral edge and posterior portion of parietal suture present; broken medially at site of large ventro-medial depression. (E) Squamosal fragment (left): only posterior portion present. (F) and (G) Squamosal fragments (left): can be rearticulated; groove for quadrate (F) and heavy ridge of bone (across F-G) against which process of exoccipital abuts present; E-F-G from same squamosal. (H) Cervical vertebra #7: end of right transverse process lost. (I) Cervical vertebra #6. Note: J, K, L, M, and T not figured. (J) Cervical vertebra #8?: 50% of vertebra recovered as float; neural arch reconstructed. (K) Two unidentified fragments: wedged against sacrum; one fragment well rounded. (L) Frill? fragment (27x10 cm) nestled against sacrum. (M) Frill? fragment (14x10 cm). (T) Turtle shell fragments scattered throughout quarry; many from one individual of *Adocus*. Also present, but not shown, are many unidentifiable bone fragments (<12 cm length).

*Stygimys*. The dinosaur remains were quarried from the basal lag of the channel fill, which also contained clay balls, fragments of turtle shell, and carbonized plant debris (V87152, Figures 6, 7). Two molars of *Ragnarok* and two incisors of *Stygimys* were collected .7-.8 m and 1.5-1.7 m above the base of the channel fill, respectively (Figure 6). These specimens occur within troughs containing clay balls and other vertebrate fossils. It is uncertain if the dinosaur remains and the mammals occur within sediments deposited in the same depositional event (see discussion of Upper Tedrow Channel below).

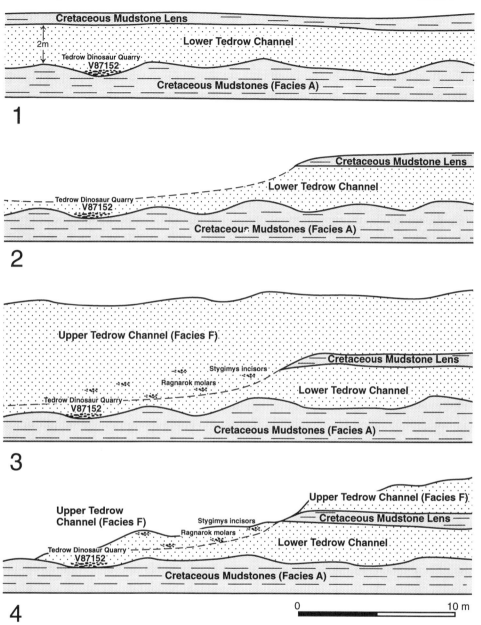

WEST                  EAST

Cretaceous Mudstone Lens

2m

Lower Tedrow Channel

Tedrow Dinosaur Quarry
V87152

Cretaceous Mudstones (Facies A)

**1**

Cretaceous Mudstone Lens

Lower Tedrow Channel

Tedrow Dinosaur Quarry
V87152

Cretaceous Mudstones (Facies A)

**2**

Upper Tedrow Channel (Facies F)

Stygimys incisors
Ragnarok molars

Cretaceous Mudstone Lens

Lower Tedrow Channel

Tedrow Dinosaur Quarry
V87152

Cretaceous Mudstones (Facies A)

**3**

Upper Tedrow
Channel (Facies F)

Stygimys incisors
Ragnarok molars

Upper Tedrow Channel (Facies F)

Cretaceous Mudstone Lens

Lower Tedrow Channel

Tedrow Dinosaur Quarry
V87152

Cretaceous Mudstones (Facies A)

0                  10 m

**4**

FIGURE 8. Scenario 1 sequence. (1) Deposition of Lower Tedrow Channel fill with dinosaur remains in basal lag, capped by mudstone lens with Cretaceous pollen. (2) Partial removal of 1 during entrenchment of Upper Tedrow Channel. Dinosaur remains in Lower Tedrow Channel fill not disturbed. (3) Deposition of Upper Tedrow Channel fill, including the mammals *Stygimys* and *Ragnarok*. (4) Area as exposed today. In this scenario, the associated dinosaur remains and the mammals were deposited in separate filling events or stories. Based on pollen, the lower story with the dinosaur bones would be Cretaceous and the higher story with the mammal teeth would be Paleocene.

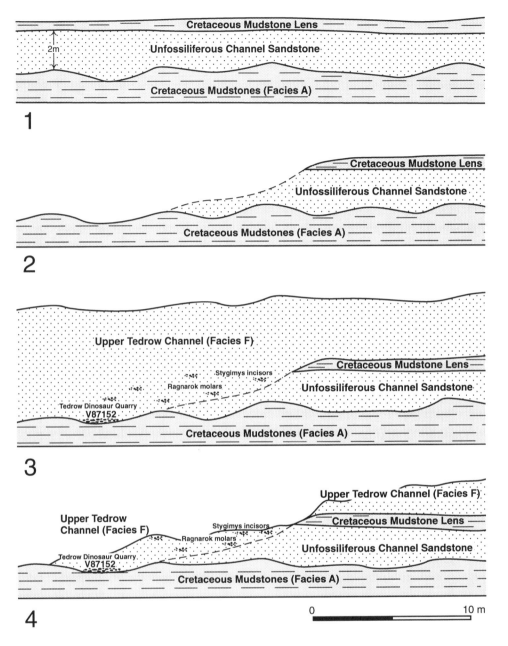

WEST                                                          EAST

**1**

**2**

**3**

**4**

FIGURE 9. Scenario 2 sequence. (1) Deposition of unfossiliferous sandstone channel fill, capped by a mudstone lens containing Cretaceous pollen. (2) Partial removal of 1 during entrenchment of Upper Tedrow Channel. (3) Deposition of Upper Tedrow Channel fill, with dinosaur remains in basal lag and the mammals *Stygimys* and *Ragnarok* higher in the fill. (4) Area as exposed today. In this scenario, all the vertebrate remains in the "Lower Tedrow" Channel are Paleocene unless reworked from Cretaceous sediments. Also, the "Lower Tedrow" Channel is a localized depression in the Upper Tedrow Channel, and both were filled at about the same time.

The "Lower Tedrow" Channel could be Cretaceous or Paleocene, depending on whether it is part of the Upper Tedrow Channel. Situated 2 m above (topographically) the base of the "Lower Tedrow" Channel is a thick mudstone lens containing Cretaceous pollen (Lofgren and Hotton, 1988), which is overlain by the Upper Tedrow Channel (Figure 6). It is uncertain whether this mudstone ever extended across the part of the channel fill that produced the dinosaur remains (contra Lofgren and Hotton, 1988) and mammal teeth. Erosion has removed the critical area in which the Cretaceous mudstone lens would have been either in superposition above part of the "Lower Tedrow" Channel or clearly cut out by the Upper Tedrow Channel (Figures 8, 9). If the latter were true, then the "Lower Tedrow" Channel would represent a localized depression in the Upper Tedrow Channel (Figure 9, nos. 2-3).

The Brown-Grey Channel contains a well sampled Bug Creek assemblage in its basal lag (V87072, Figure 6) and its palynological age is Paleocene. This channel fill cuts the mudstone lens containing Cretaceous pollen. Lofgren and Hotton (1988) thought that the mudstone lens overlay the Brown-Grey Channel; this is now known to be incorrect.

Upper Tedrow is a multi-storied channel fill whose lowest story may include part or all of the "Lower Tedrow" Channel. The Upper Tedrow Channel is Paleocene because it cuts the Brown-Grey (Figure 6; Section GG, Plate 3) and Second Level channels (Sections P, 856, Plate 4), both of which produce Paleocene pollen (Table 3). Therefore, if part of the "Lower Tedrow" Channel is a localized depression of the Upper Tedrow Channel, it is probably Paleocene in age.

On the basis of preliminary palynological and geological investigations outlined in Lofgren and Hotton (1988), it was thought that the sequence of events in the Tedrow Area was: (1) deposition of the Brown-Grey Channel; (2) entrenchment of the "Lower Tedrow" Channel containing the dinosaur remains and subsequent filling; (3) deposition of the mudstone lens overlying "Lower Tedrow" during the Cretaceous; (4) entrenchment of Upper Tedrow into all three of the above. If this sequence of events were correct, then the "Lower Tedrow" and Brown-Grey channels would be Cretaceous and the Upper Tedrow Channel would be Paleocene. Trenching in the Tedrow area in 1989 revealed that the Brown-Grey Channel cuts the Cretaceous mudstone lens. This is corroborated by the presence of a Paleocene pollen assemblage in samples 88DLL7-14-13 and 88DLL7-14-30 from the Brown-Grey Channel, which also indicates that this channel is younger than the Cretaceous mudstone lens. However, the "Lower Tedrow" Channel appears to cut the Brown-Grey Channel when traced westward. This is impossible, because the Brown-Grey Channel cannot be cut by a channel ("Lower Tedrow") that is capped by a unit (mudstone lens) cut by the Brown-Grey Channel (Figure 6). Therefore, part of the "Lower Tedrow" Channel must be equivalent to the Upper Tedrow scouring event (that cut the Brown-Grey Channel and the Cretaceous mudstone lens). This fact negates a two-stage channeling hypothesis in which all vertebrate remains in the "Lower Tedrow" Channel are Cretaceous, and the "Lower" and Upper Tedrow channels are two distinct cut-and-fill events (contra Lofgren and Hotton, 1988).

Based on available geologic data from the Tedrow Area, two possible scenarios of depositional events are presented in Figures 8 and 9, and neither can be falsified by direct field observation. If correct, the first hypothesis (Figure 8) would indicate a Cretaceous age for the dinosaur remains and a Paleocene age for the mammals. The second (Figure 9) would indicate a Paleocene age for both the dinosaur remains and the mammals, unless reworking of fossils occurred.

If the dinosaur remains were not reworked, then choosing between the two hypotheses has important implications for the timing of dinosaur extinction. The first hypothesis is more in line with conventional palynological-faunal correlations in which dinosaur remains indicate a Cretaceous age and the mammal *Ragnarok* indicates a Paleocene age. *Ragnarok* is only known to occur in clear association with Paleocene palynofloras at McGuire Creek and elsewhere in McCone County (Sloan et al., 1986; Rigby et al., 1987; Rigby, 1989). The associated ceratopsian remains would indicate a Cretaceous age for the base of the channel fill.

The second hypothesis is more controversial, because the presence of associated dinosaur remains in a Paleocene channel fill might suggest dinosaur survival into the Paleocene. The association of disarticulated skeletal material (Figure 7) might suggest that the bones were not reworked. However, reworking of dinosaur remains that persist in near association would be possible if the remains were eroded from nearby bank material and underwent little or no transport. Dinosaur remains have been found in channel deposits containing Paleocene pollen at McGuire Creek and elsewhere (Sloan et al., 1986; Rigby et al., 1987) but that these remains are not reworked from Cretaceous sediments has not been satisfactorily documented (Lofgren et al., 1990). Therefore, if the second hypothesis is correct, the dinosaur remains probably were reworked from nearby Cretaceous sediments into which the Upper Tedrow Channel was entrenched.

In summary, difficulties encountered in biostratigraphic study of complex fluvial deposits in the Tedrow Area indicate that determining a clear sequence of depositional events in even a limited area can be difficult. The complexity of channel cutting and filling in the upper Hell Creek Formation is imposing.

# REWORKING OF FOSSILS

The most perplexing problem concerning vertebrate biostratigraphy of the K-T transition in eastern Montana is that of reworked fossils. As the data presented in the chapter above on Geological and Palynological Correlations show, the potential for reworking in the upper Hell Creek Formation is staggering. Channel fills containing Bug Creek assemblages were deeply entrenched into Lancian strata (up to 20 m), and a tremendous volume of sediment was eroded (see Figure 5). Exhumation and reburial of dinosaur and Lancian mammal fossils during the cutting and filling of these channels must have occurred, but to what extent? This question is especially critical in the uppermost Hell Creek Formation where there is a significant change in the vertebrate fauna preserved within a series of channel fills, as evidenced by the presence of Lancian, Bug Creek, and Puercan assemblages.

Originally, the Bug Creek assemblages (Bug Creek Anthills, Bug Creek West, and Harbicht Hill) were interpreted as being composed of the remains of animals that once lived contemporaneously (Sloan and Van Valen, 1965). Although these fossils were recovered from large channel fills, the possibility that some of them might be reworked was not considered, at least in print. Later, the depositional setting of these vertebrate remains and the potential for reworked fossils were more thoroughly appreciated (Smit and Van der Kaars, 1984; Bryant et al., 1986; Retallack et al., 1987; Fastovsky, 1987; Eaton et al., 1989; Bryant, 1989; Lofgren et al., 1990; Lofgren and Hotton, 1991a, 1991b).

The confusion over which fossils are reworked impinges on K-T correlations because many channels that produce dinosaurs and potentially reworked Lancian mammals yield Paleocene palynofloras. Some researchers argue that dinosaurs (but not Lancian mammals?) persisted into the Paleocene (Sloan et al., 1986; Rigby et al., 1987; Rigby, 1987, 1989), while others suggest that these remains are reworked (Smit and Van der Kaars, 1984; Bryant et al., 1986; Retallack et al., 1987; Bryant, 1989). At issue is whether Bug Creek assemblages represent (1) the remains of animals that lived contemporaneously, or (2) the remains of animals from temporally distinct assemblages that have been mixed through reworking.

Table 4. Mammalian species and dinosaur remains present in local faunas in eastern Montana

| LOCAL FAUNA: | FC | BA[a] | BW | HH[a] | LR | SL | BS | BG | SR | UU | HE | ZL |
|---|---|---|---|---|---|---|---|---|---|---|---|---|
| *DINOSAUR REMAINS | X | X | X | X | X | X | X | X | X | X | R | |
| **MULTITUBERCULATA** | | | | | | | | | | | | |
| *Meniscoessus robustus | X | X | X | X | X | X | X | X | X | X | | |
| *Essonodon browni | X | X | | | | X | | | | | | |
| *Cimolodon nitidus | X | X | X | | | | | X | | | | |
| *Cimolomys gracilis | X | X | | | | | | | | | | |
| *Paracimexomys priscus | X | | | | | | | | | | | |
| ?Neoplagiaulax burgessi | X | | | | | | | | | | | |
| Mesodma spp. | X | X | X | X | X | X | X | X | X | X | X | X |
| Stygimys kuszmauli | | X | X | X | X | X | X | X | X | X | | |
| Cimexomys gratus | | X | ? | X | X | X | X | X | X | X | | |
| Cimexomys minor | X | X | X | X | X | X | X | X | X | X | X | |
| Catopsalis alexanderi | | | | X | | | | | | | X | X |
| Catopsalis joyneri | | X | X | X | | | X | | X | | | |
| Catopsalis sp. indet. | | | | | | | X | X | X | X | | |
| Acheronodon garbani | | | | | | | | | | | X | |
| **MARSUPIALIA** | | | | | | | | | | | | |
| *Pediomys hatcheri | X | X | X | | X | X | | X | | X | | |
| *Pediomys krejcii | X | X | ? | | | X | | X | X | X | | |
| *Pediomys florencae | | X | ? | | | X | | X | X | X | | |
| *Pediomys elegans | X | X | X | | X | X | | X | X | X | | |
| *Pediomys cooki | cf | X | ? | | | | | | | | | |
| *Pediomys sp. indet. | | | | X | | | | X | | X | | |
| *Protalphadon lulli | X | | | | | | | | | | | |
| *Alphadon "wilsoni" | X | X | | cf | | X | | X | X | X | | |
| *Alphadon rhaister | X | | | | X | X | | X | X | X | | |
| *Alphadon sp. indet. | | | | | X | | | X | | X | | |
| *Alphadon marshi | X | X | | | | | | X | | | | |
| *Glasbius twitchelli | X | X | X | cf | | | | X | X | X | | |
| *Didelphodon vorax | | X | X | | X | X | | X | X | X | | |
| Peradectes cf pusillus | | | | | | | | X | X | X | X | X |
| **EUTHERIA** | | | | | | | | | | | | |
| *Gypsonictops spp. | X | X | X | X | X | X | | X | X | X | | |
| *Batodon tenuis | X | X | | | | X | | | X | X | | |
| Cimolestes spp. | X | X | | | | | | | | | | |
| Procerberus formicarum | | X | X | X | X | X | X | X | X | X | X | X |
| **"CONDYLARTHRA"** | | | | | | | | | | | | |
| Protungulatum donnae | | X | X | X | X | X | X | X | X | X | cf | |
| Protungulatum gorgun | | X | X | | X | X | X | X | X | X | | |
| Mimatuta morgoth | | X | X | X | X | X | X | X | X | X | X | |
| Mimatuta minuial | | | | | | X | | X | | X | | |
| Mimatuta sp. indet. | | | | | | | X | X | X | X | | X |
| Ragnarok nordicum | | | X | | X | X | X | X | X | X | ? | |
| Ragnarok engdahli | | | | | | | | | | | X | |
| Oxyprimus erikseni | | X | | X | X | X | | X | X | X | X | |

a. Presence of *Catopsalis alexanderi* at HH and *Cimexomys gratus* at BA and HH based on study of UCMP specimens (see Appendix 1). See Table 8 for sources of taxonomic data and abbreviation key.

*: Lancian taxa considered here as reworked.

R: rare.

cf: tentative referral.

?: questionable record.

---

## IDENTIFICATION OF REWORKED FOSSILS

The problem then becomes one of identifying, if possible, which taxa have been reworked and to what extent reworking has extended their geologic range. Lancian mammal taxa whose presence in Bug Creek assemblages is probably the result of reworking, because they only occur in channel fills (many of which yielded Paleocene palynofloras) that are entrenched into Lancian sediments, are indicated by asterisks in Table 4. These taxa are identified as probably reworked. In all subsequent discussion of reworking, the term "Lancian mammals" will refer to those denoted by asterisks in Table 4.

To aid in recognition of reworked fossils, transport experiments on crocodile and dinosaur teeth were conducted, with mixed results (Argast et al., 1987; Ely and Rigby, 1989). These experiments compared laboratory controlled abrasion of fossil and extant vertebrate teeth with the abrasion evident on fossil teeth recovered from channel deposits (i.e., no abrasion=low potential of having been reworked; abraded=high potential of having been reworked).

Before reworking and/or transporting of vertebrate remains can be discussed further, the terminology employed needs to be clarified. Transported refers here to animal remains that are moved from one place to another, reworked here means to reprocess animal remains, in this case, those that were deposited, exhumed, and then reburied. Therefore, two very different components are involved: a spatial (transported: to move some distance) and a temporal component (reworked: to reprocess).

All fossils in channel deposits are transported to some degree, but not all fossils are reworked. Therefore, the amount of transport and the potential that the remains were reprocessed are interrelated only when reworked remains have undergone enough transport so that abrasion is noticeable. Elements that are transported a short distance usually will not be abraded, whether they are reworked or not (unless bone at the surface of a channel floor is sand-blasted). Conversely, elements that were transported and abraded before initial burial may not be reworked. Therefore, it is guesswork to link abrasion or absence of abrasion with reworking. Transport experiments on vertebrate remains are helpful as indicators of how to recognize a fossil that has been transported (Ely and Rigby, 1989; but see Argast et al., 1987), but they do not provide the insight necessary to reconstruct the taphonomic history of any particular fossil with confidence.

To illustrate this point, the basal lag of the Black Spring Coulee Channel (V87030, Section II, Plate 4) yields many well-preserved dinosaur elements which are dispersed within a conglomerate composed of cobble and boulder-sized angular blocks of mudstone. Some dinosaur bones are still encased in mudstone blocks, and one of these yielded a Cretaceous palynoflora (Lofgren et al., 1990). In contrast, a thin siltstone lens within this channel fill yielded a Paleocene palynoflora. Therefore, based on sedimentological and palynological analysis, these dinosaur remains were reworked from Cretaceous floodplain deposits into a Paleocene channel fill (Lofgren et al., 1990). This example demonstrates the potential difficulty in distinguishing reworked material from vertebrate remains that were initially buried during channel filling. Another example of the reworking of Cretaceous fossils into Tertiary deposits is provided by Eaton et al. (1989).

## "PALEOCENE" DINOSAURS

Geological and faunal data from Bug Creek West, Harbicht Hill, Ferguson Ranch, Doverzee, and other sites in McCone County, eastern Montana, have been presented in support of the hypothesis that dinosaurs survived into the Paleocene (Sloan et al., 1986; Rigby et al., 1987; Rigby, 1987, 1989). Unabraded, isolated teeth of dinosaurs from Paleocene channel fills (containing Paleocene palynofloras or truncating beds of Paleocene age) have been interpreted as unreworked and cited as evidence in support of this assertion. However, these records are limited to channel deposits in the uppermost Hell Creek Formation that are deeply incised into Cretaceous dinosaur-bearing strata (see Rigby et al., 1987: figures 3, 5). In overbank stratigraphic sequences uncomplicated by major channeling events, the highest dinosaur remains occur approximately 2m below the palynological K-T boundary (Archibald, 1987c). If large dinosaur bones had been reworked into similar channel facies at McGuire Creek (Lofgren et al., 1990), would records of "Paleocene" dinosaurs based upon isolated teeth bear serious consideration? J.K. Rigby, Jr. (1987, 1989) provides the most complete argument favoring the existence of "Paleocene" dinosaurs. His points favoring this interpretation, here numbered 1-5, are evaluated individually.

1. Clay-pebble clasts and vertebrate remains in channel lag deposits probably would not survive lengthy transport; therefore, the presence of clasts associated with dinosaur remains indicates that neither was reworked (Rigby, 1987). *Evaluation*: If the clay clasts and vertebrate remains are reworked and not transported significant distances, this argument is nullified. Rigby argues that clay pebbles will not survive significant transport and therefore must be locally derived. If these clay clasts were eroded from bank material, the fact that they are preserved at all indicates little transport. This would also apply to the vertebrate remains eroded from the same bank material from which the clay pebbles were derived.

2. The potential source of reworked fossils is overbank deposits, which are not very fossiliferous and therefore are not likely to be the source of the extraordinary abundance of vertebrates seen in screen-washed concentrates (Rigby, 1987). *Evaluation*: Eroded overbank facies are the source of the abundant reworked dinosaur remains in one channel fill analyzed by Lofgren et al. (1990). Rigby apparently made the assumption that eroded overbank deposits are the primary source for reworked dinosaur remains. In addition, erosion of pre-existing channel deposits, which are com-

mon in the upper Hell Creek Formation, would have been a major source for reworked fossils. Finally, the preservation potential of animal remains deposited in overbank and channel facies might be quite different. Diagenetic factors, especially soil formation, might preferentially destroy vertebrate remains in overbank facies either prior to, or after deep burial.

3. Many "Paleocene" dinosaur remains appear unabraded, and this would not be the case had these remains undergone transport and reworking (Rigby, 1987). *Evaluation*: As discussed above, fossils that are reworked and undergo little transport may not exhibit abrasion. Absence of abrasion does not exclude the possibility of reworking.

4. Many Lancian mammals do not appear associated with "Paleocene" dinosaur remains; if reworking occurred it must have been selective, which is improbable (Rigby, 1987). *Evaluation*: Bug Creek sites from McGuire Creek yielding Paleocene palynofloras contain dinosaurian remains associated with a diverse fauna of Lancian mammals (compare BG to HH and BA, Table 4). The Bug Creek West and Harbicht Hill sites, which yield Paleocene palynofloras and "Paleocene" dinosaurs (Sloan et al., 1986; Rigby et al., 1987), are not thoroughly sampled, nor are their vertebrate faunas completely described. If these sites were resampled and their faunas restudied, many Lancian taxa would probably be discovered. Many Paleocene McGuire Creek local faunas contain Lancian mammals as well as dinosaur remains. If dinosaur remains in Paleocene channel fills were not all reworked, it would be equally likely that Lancian mammals in Paleocene channel fills at McGuire Creek were not all reworked (contra Sloan et al., 1986; Rigby et al., 1987; Rigby, 1989).

Also, there are in McGuire Creek local faunas at V87072 and V87035-8 higher relative abundances of Lancian mammals than in Bug Creek Anthills (Table 5), which is presumably older (see the next chapter, on McGuire Creek Biochronology and the "Bugcreekian Age"). On the basis of arguments similar to those claiming the existence of "Paleocene" dinosaurs, it would seem that Lancian mammals also survived into the Paleocene because they become more abundant in younger channel fills, from Bug Creek Anthills (older) to V87072 and V87035-8 (younger). However, the reworking of vertebrate remains does not follow any predictable pattern, and the amount of reworked material will depend on the amount of previously deposited sediment eroded by the stream as well as the fossil content of the eroded sediment. These factors vary independently from the relative age of a channel fill.

5. The base of the Doverzee Channel fill lies just above the upper "Z" lignite (palynologically Paleocene) and yields dinosaur teeth but lacks Lancian mammals (Rigby, 1989). Therefore, the Doverzee Channel was never entrenched into Cretaceous sediment, and the dinosaur teeth in the channel fill are not reworked from Cretaceous beds. *Evaluation*: Independent mapping of the Doverzee channel fill (SW quadrant of Section 16, T 22 N, R 43 E) revealed the following points: (A) When the Doverzee Channel fill is traced southeast (downstream) across discontinuous exposures (based on paleocurrent azimuths and lithologic similarity) from the spot where the dinosaur teeth were reportedly collected (Rigby, 1989), it cuts out the upper "Z" lignite (Section OO, Plates 1 and 4). (B) Traced further southeast, the Doverzee Channel cuts into strata many meters below the "Z" lignite to a level presumably Lancian in age (contra Rigby,

Table 5. Relative abundance (percentage of total sample) of mammalian genera from selected screenwash sites

| Locality:<br>(NISP)[e]: | V87072[a]<br>(456) | V87035-38[a]<br>(954) | V73087[b]<br>(213) | V74111[b]<br>(502) | BA[c]<br>(6000) | BA[d]<br>(1665) |
|---|---|---|---|---|---|---|
| **MULTITUBERCULATA** | | | | | | |
| *Mesodma* | 73.0 | 67.0 | 31.0 | 45.0 | 43.0 | 59.0 |
| *Stygimys* | 6.8 | 12.0 | | 6.6 | 26.0 | 13.0 |
| *Cimexomys* | 7.4 | 5.9 | | 8.0 | 1.1 | 3.0 |
| *\*Cimolomys* | | | .9 | | .1 | |
| *\*Cimolodon* | .2 | | 6.1 | | .1 | |
| *\*Meniscoessus* | .2 | .6 | .5 | | .9 | .2 |
| *\*Essonodon* | | | 1.0 | | .1 | |
| *\*Paracimexomys* | | | 7.0 | | | |
| *Catopsalis* | | .1 | | .2 | 3.9 | 3.2 |
| *?Neoplagiaulax* | | | 1.9 | | | |
| *Acheronodon* | | | | .2 | | |
| **MARSUPIALIA** | | | | | | |
| *\*Pediomys* | 1.7 | 1.5 | 4.7 | | 1.0 | .9 |
| *\*Alphadon* | 1.5 | .8 | 28.0 | | .02 | .4 |
| *\*Didelphodon* | | .3 | | | .1 | .1 |
| *\*Glasbius* | .9 | 1.3 | 12.0 | | .1 | |
| *Peradectes* | .4 | .4 | | 14.0 | | |
| **EUTHERIA** | | | | | | |
| *\*Gypsonictops* | .7 | .6 | 3.8 | | .6 | .4 |
| *\*Batodon* | | .2 | 1.9 | | | |
| *Cimolestes* | | | 2.3 | | .1 | |
| *Procerberus* | 2.3 | 5.0 | | 21.0 | 7.1 | 6.7 |
| **"CONDYLARTHRA"** | | | | | | |
| *Protungulatum* | .7 | 2.0 | | 1.2 | 17.0 | 12.0 |
| *Mimatuta* | 1.1 | 1.5 | | 2.6 | | |
| *Ragnarok* | .9 | .9 | | .2 | | |
| *Oxyprimus* | .2 | .5 | | 2.0 | | |
| TOTAL *LANCIAN: | 5.2% | 5.3% | 65.9% | 0.0% | 3.0% | 2.0% |

a. McGuire Creek sites: V87072= part of Brown-Grey Local Fauna, and V87035-38=Up-Up-the-Creek Local Fauna.
b. V73087= Flat Creek 5, V74111= Worm Coulee 1, data from Archibald, 1982.
c. Bug Creek Anthills data from Sloan and Van Valen, 1965.
d. Bug Creek Anthills data from Estes and Berberian, 1970.
e. Number of identifiable specimens per taxon at each site.
*: Lancian genera not found in Paleocene sites other than large channel fills and considered here as reworked (except for occurrences at V73087).

1989). (C) Although southeast is the direction of paleoflow, if the Doverzee stream cut through the upper "Z" lignite downstream, it is probable (but not demonstrable) that this occurred upstream as well. Therefore, eroded Lancian strata could easily have been the source of the dinosaur teeth. (D) Given the history of claims that the absence of taxa is an artifact of small sample size (e.g., at Bug Creek West and Harbicht Hill), the claimed absence of Lancian mammals at the Doverzee locality cannot be evaluated until data on sample size and methods of collecting are given.

In summary, Rigby's (1987, 1989) arguments fall far short of substantiating the claim of dinosaur survival into the Paleocene.

## BUG CREEK FAUNAS REPRESENT CONTEMPORANEOUS ASSEMBLAGES

Based on data from McGuire Creek, arguments similar to those supporting "Paleocene" dinosaurs could be proposed to support the hypothesis that Bug Creek faunas each represent contemporaneous assemblages. Pertinent evidence is as follows:

1. The excellent condition of Lancian mammal remains, especially those of marsupials, is evidence that these fossils were not all reworked. Three jaw and maxillary fragments containing 1-3 teeth of *Pediomys* and *Glasbius* were discovered at the Up-Up-the-Creek and Brown-Grey Channel localities. These are very delicate fossils and presumably would have been destroyed easily during erosion and redeposition by stream processes. *Evaluation*: These remains are found in a stream deposit and must have undergone some transport. Therefore, it is possible that they were reworked and underwent minimal transport before reburial. This argument would also apply to the presence of well-preserved dinosaur bones and teeth in Paleocene channels. In fact, it has been documented that well-preserved dinosaur remains were reworked into a Paleocene channel fill (Lofgren et al., 1990).

2. Lancian mammal and dinosaur remains are relatively abundant at certain McGuire Creek sites, such as V87072 (Brown-Grey Channel) and V87035-8 (Up-Up-the-Creek Local Fauna), more so than might be expected if reworking were the cause. In the case of the mammals, the marsupials *Glasbius twitchelli* and *Pediomys elegans* are especially abundant. *Pediomys elegans* is more abundant than any "condylarth" species at V87072, as is *Glasbius twitchelli* at V87035-8 (Table 6). Also, Lancian mammal taxa (denoted with * in Tables 5, 6, and 7) constitute over 5% of the total mammalian fauna in NISP (Number of Identifiable Specimens per Taxon) at both the generic and specific levels, and 16-19% in MNI (Minimum Number of Individuals) at the specific level from these two localities (Tables 6, 7). In a channel-fill depositional setting, relative abundance data are unlikely to reflect the relative abundances of animals that lived in the area when the deposit was formed. However, can the reworking hypothesis explain these apparently high relative abundances of Lancian taxa, especially when MNIs are considered? *Evaluation*: Vertebrate fossils in channel deposits are hydraulically sorted and transported, so elements identified as coming from one taxon have a low probability of being from one individual. Therefore, it is more appropriate in this deposi-

tional setting to use NISPs than MNIs (Badgley, 1986), because MNIs will overestimate the relative abundance of rare taxa. Flat Creek, the stratigraphically highest Lancian local fauna in the Hell Creek Formation (Figure 1b), produces about 66% Lancian taxa (NISP) at the generic level (Table 5). At McGuire Creek sites V87072 and V87035-8, the relative abundance of Lancian mammals is much lower (5%, NISP), which would be expected if reworking were responsible for their presence. Lancian mammal remains in McGuire Creek channel fills probably are fossils redeposited following erosion from facies containing abundant Lancian mammals.

The two arguments presented above are based on robust data from McGuire Creek, but fall short of demonstrating the contemporaneity of the components of Bug Creek assemblages (Lancian-Puercan mammals and dinosaurs).

## BUG CREEK FAUNAS CONTAIN REWORKED DINOSAURS/LANCIAN MAMMALS

Evidence suggesting that association of dinosaurs and Lancian mammals in Bug Creek assemblages at McGuire Creek is the result of reworking is outlined below:

1. Bug Creek assemblages from McGuire Creek are restricted to the basal lags of channels which were incised into Lancian strata (up to 20 m) (see chapter above on Geological and Palynological Correlations). This correlation of channel incision with the presence of Lancian mammals and dinosaurs in Bug Creek assemblages is not coincidental. Such a particular mixture of faunal components was caused by reworking. Similar arguments were proposed by Smit and Van der Kaars, 1984; Bryant et al., 1986; Retallack et al., 1987; and Bryant, 1989, based on the depositional setting of the original Bug Creek sites.

2. The reworking hypothesis can be tested with analyses of Facies G strata that overlie (or are laterally equivalent to) channel fills containing Bug Creek assemblages in the upper Hell Creek Formation at McGuire Creek. Facies G strata represent a low-energy, floodplain depositional setting where the possibility of reworking can be virtually eliminated. Unfortunately, this depositional setting usually is not conducive to producing microvertebrate concentrations. However, dinosaur remains commonly are present in floodplain deposits, and presumably would be present in Facies G floodplain deposits that are equivalent in age to channel facies that contain dinosaur remains. However, dinosaurs were not recovered from Facies G deposits. The absence of dinosaur remains in Facies G floodplains (which yield other vertebrates) is especially significant for laterally accreted channel facies (Facies E), where floodplain deposits and contemporaneous channel facies are juxtaposed (Figure 10). One would expect to find dinosaur remains in floodplain sediments (Facies G) if dinosaurs had been extant during the deposition of directly adjacent channel fills (Facies E). Because dinosaur remains occur in the basal lags of channels but are absent in age-equivalent floodplain deposits, their presence in these channels is probably the result of reworking.

3. Channel fills with nearly identical assemblages (those taxa not marked by * in

Table 6. Relative abundance (percentage of total sample) of mammalian species from selected screenwash sites.

| LOCALITY: | V87072[a] | V87072[a] | V87035-8[a] | V87035-8[a] | BA[b] |
|---|---|---|---|---|---|
| | NISP (456) | MNI (73) | NISP (954) | MNI (184) | NISP (6000) |
| **MULTITUBERCULATA** | | | | | |
| Mesodma spp. | (345) 73.0 | (38) 52.0 | (639) 67.0 | (82) 45.0 | 43.0 |
| Stygimys kuszmauli | (32) 6.8 | (5) 6.8 | (109) 12.0 | (21) 11.4 | 25.7 |
| Cimexomys gratus | (25) 5.3 | (5) 6.8 | (39) 4.1 | (10) 5.4 | present |
| Cimexomys minor | (10) 2.1 | (3) 4.1 | (17) 1.8 | (6) 3.3 | 1.1 |
| Catopsalis alexanderi | | | (1) .1 | .5 | |
| Catopsalis joyneri | | | | | 3.9 |
| *Meniscoessus robustus | (1) .2 | 1.4 | (6) .6 | (3) 1.6 | .9 |
| *Essonodon browni | | | | | .1 |
| *Cimolodon nitidus | | | | | .1 |
| *Cimolomys gracilis | | | | | .1 |
| **MARSUPIALIA** | | | | | |
| *Pediomys hatcheri | | | (2) .2 | (2) 1.1 | .2 |
| *Pediomys krejcii | (1) .2 | 1.4 | (3) .3 | (2) 1.1 | |
| *Pediomys florencae | (1) .2 | 1.4 | (3) .3 | (2) 1.1 | |
| *Pediomys elegans | (6) 1.3 | (4) 5.5 | (3) .3 | (2) 1.1 | .5 |
| *Pediomys sp. indet. | (1) .2 | 1.4 | (3) .3 | (2) 1.1 | .3 |
| *Alphadon "wilsoni" | (4) .9 | (1) 1.4 | (1) .1 | .5 | |
| *Alphadon rhaister | (1) .2 | 1.4 | (1) .1 | .5 | |
| *Alphadon sp. indet. | (1) .2 | 1.4 | (5) .5 | (3) 1.6 | |
| *Alphadon marshi | (1) .2 | 1.4 | | | .02 |
| *Glasbius twitchelli | (3) .7 | (1) 1.4 | (12) 1.3 | (6) 3.3 | .1 |
| *Didelphodon vorax | | | (3) .3 | (1) .5 | .1 |
| Peradectes cf pusillus | (2) .4 | (1) 1.4 | (4) .4 | (3) 1.6 | |
| **EUTHERIA** | | | | | |
| *Gypsonictops illuminatus | (3) .7 | (1) 1.4 | (6) .6 | (3) 1.6 | .6 |
| *Batodon tenuis | | | (2) .2 | (1) .5 | |
| Cimolestes incisus | | | | | .1 |
| Procerberus formicarum | (11) 2.3 | (2) 2.7 | (47) 5.0 | (7) 3.8 | 7.1 |
| **"CONDYLARTHRA"** | | | | | |
| Protungulatum donnae | (3) .7 | (2) 2.7 | (11) 1.2 | (6) 3.3 | 16.7 |
| Protungulatum gorgun | | | (8) .8 | (6) 3.3 | |
| Mimatuta morgoth | | | (3) .3 | (3) 1.6 | |
| Mimatuta minuial | (1) .2 | 1.4 | (2) .2 | (2) 1.1 | |
| Mimatuta sp. indet. | (1) .2 | 1.4 | (9) 1.0 | (4) 2.2 | |
| Ragnarok nordicum | (4) .9 | (2) 2.7 | (8) .8 | (5) 2.7 | |
| Oxyprimus erikseni | (1) .2 | 1.4 | (5) .5 | (3) 1.6 | |
| TOTAL *LANCIAN: | 5.0% | 19.5% | 5.1% | 15.6% | 3.1% |

a. McGuire Creek sites: V87072= part of Brown-Grey Channel Local Fauna, and V87035-38= Up-Up-the-Creek Local Fauna.
b. Bug Creek Anthills data from Sloan and Van Valen, 1965.
*. Lancian species considered here as reworked.

Table 7. Relative abundance (percentage of total sample) of mammalian species from selected screenwash sites

| LOCALITY:<br>NISP: | V87072[a]<br>(456) | V87035-8[a]<br>(954) | V73087[b]<br>(213) | V74111[b]<br>(502) | BA[c]<br>(6000) |
|---|---|---|---|---|---|
| MULTITUBERCULATA | | | | | |
| Mesodma spp. | 73.0 | 67.0 | 31.0 | 45.0 | 43.0 |
| Stygimys kuszmauli | 6.8 | 12.0 | | 6.6 | 25.7 |
| Cimexomys gratus | 5.3 | 4.1 | | 6.2 | present |
| Cimexomys minor | 2.1 | 1.8 | | 1.8 | 1.1 |
| Catopsalis alexanderi | | .1 | | .2 | |
| Catopsalis joyneri | | | | | 3.9 |
| *Meniscoessus robustus | .2 | .6 | .5 | | .9 |
| *Essonodon browni | | | 1.0 | | .1 |
| *Cimolodon nitidus | | | 6.1 | | .1 |
| *Cimolomys gracilis | | | 1.0 | | .1 |
| *Paracimexomys priscus | | | 7.0 | | |
| ?Neoplagiaulax burgessi | | | 1.9 | | |
| Acheronodon garbani | | | | .2 | |
| MARSUPIALIA | | | | | |
| *Pediomys hatcheri | | .2 | .5 | | .2 |
| *Pediomys krejcii | .2 | .3 | 2.4 | | present |
| *Pediomys florencae | .2 | .3 | | | present |
| *Pediomys elegans | 1.3 | .3 | 1.4 | | .5 |
| *Pediomys cf cooki | | | .5 | | present |
| *Pediomys sp. indet. | | .3 | | | .3 |
| *Protalphadon lulli | | | 1.4 | | |
| *Alphadon "wilsoni" | .9 | .1 | 11.0 | present | |
| *Alphadon rhaister | .2 | .1 | .5 | | |
| *Alphadon sp. indet. | .2 | .5 | | | |
| *Alphadon marshi | .2 | | 15.0 | | .02 |
| *Glasbius twitchelli | .9 | 1.3 | 12.0 | | .1 |
| *Didelphodon vorax | | .3 | | | .1 |
| Peradectes cf pusillus | .4 | .4 | | 14.0 | |
| EUTHERIA | | | | | |
| *Gypsonictops illuminatus | .7 | .6 | 3.8 | | .6 |
| *Batodon tenuis | | .2 | 1.9 | | |
| Cimolestes spp. | | | 2.3 | | .1 |
| Procerberus formicarum | 2.3 | 5.0 | | 21.0 | 7.1 |
| "CONDYLARTHRA" | | | | | |
| Protungulatum donnae | .7 | 1.2 | | | 16.7 |
| Protungulatum cf donnae | | | | 1.2 | |
| Protungulatum gorgun | | .8 | | | |
| Mimatuta morgoth | | .3 | | 2.6 | |
| Mimatuta minuial | .9 | .2 | | | |
| Mimatuta sp. indet. | .2 | 1.0 | | | |
| Ragnarok nordicum | .9 | .8 | | | |
| ?Ragnarok engdahli | | | | .2 | |
| Oxyprimus erikseni | .2 | .5 | | 2.0 | |

a. McGuire Creek sites: V87072= part of Brown-Grey Channel Local Fauna, and V87035-38= Up-Up-the-Creek Local Fauna.

b. V73087= Flat Creek 5, V74111= Worm Coulee 1, data from Archibald, 1982.

c. Bug Creek Anthills data from Sloan and Van Valen, 1965.

*. Lancian species considered as reworked.

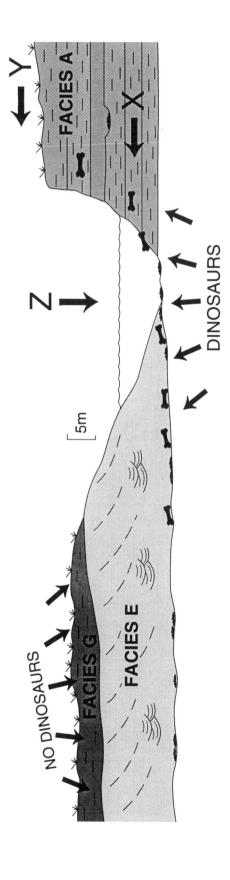

FIGURE 10. Taphonomic model adapted from Behrensmeyer, 1982, for occurrence of dinosaur remains in Paleocene (or Pu0-Pu1) Facies E channel fills at McGuire Creek. The three possible sources of the dinosaur remains (or Lancian mammals) are: (X) reworked from older strata through bank erosion, (Y) overland transport from the floodplain surface into the channel, (Z) individuals that died in the channel. Dinosaur remains are found in the basal lags of Facies E channels but not in age-equivalent floodplain deposits. The absence of dinosaur remains in floodplain deposits which aggraded juxtaposed to laterally accreting channels yielding dinosaurs suggests that the source of the remains is pathway X. If dinosaur bones and teeth were entering the channel via pathways Y or Z, their remains would be expected to occur in both the channel and floodplain deposits. Therefore, it is unlikely that dinosaurs lived during the formation of these channel fills.

Table 4) sometimes produce dinosaur remains and Lancian mammals (BG, SR, UU) and sometimes do not (HE, ZL). Based on taxa which make their first appearance in the Puercan, it was concluded that these channel fills are approximately the same age but differ primarily in their Lancian components. Lancian taxa are found only in those channels that cut Lancian strata (Table 4, compare BG, SR, and UC with HE or ZL). These data indicate that the Lancian fossils were reworked.

The question then becomes one of deciding which argument is more compelling: reworked or not reworked? I suggest that the reworking argument is more compelling for the following reasons: (1) deep incision of channels (with Bug Creek assemblages) resulted in the subsequent erosion of a tremendous volume of Lancian strata; (2) exclusive association of Bug Creek assemblages with channels that demonstrably cut Lancian strata; (3) the presence of dinosaur remains in the basal lags of channels but their absence in age-equivalent floodplain deposits that yield other vertebrates; (4) the lack of dinosaurs in the stratigraphically lowest Paleocene floodplain deposits identified palynologically; and (5) the unequivocal occurrence of Cretaceous dinosaur remains in a Paleocene channel fill (Lofgren et al., 1990).

These data are hard to dismiss, especially considering the enormous potential for reworking in this depositional and stratigraphic setting. The burden then rests on those who argue for faunal contemporanity of dinosaurs and Lancian-Puercan mammals in Bug Creek assemblages, or "Paleocene" dinosaurs, to demonstrate that the fossils in question are not reworked.

To this end, a standard for recognition of "Paleocene" dinosaurs, consisting of four criteria, was proposed: (1) articulated dinosaur remains from a Paleocene channel fill; (2) in situ dinosaur from a Paleocene floodplain; (3) abundant dinosaur remains from a channel fill that did not scour Cretaceous sediments; and (4) dinosaur remains demonstrably reworked from Paleocene sediments (Lofgren et al., 1990). These criteria provide a standard for assessing the significance of Lancian mammal and dinosaur fossils in Paleocene rocks. Until at least one of these criteria is met, the reworking hypothesis should be dismissed as unsubstantiated.

In conclusion, the presence of dinosaurs and Lancian mammals in Bug Creek assemblages (and Paleocene channels) from the upper Hell Creek Formation at McGuire Creek is a result of reworking. Similarly, the claim of dinosaur survival into the Paleocene is rejected.

# MCGUIRE CREEK BIOCHRONOLOGY
# AND THE "BUGCREEKIAN AGE"

A "Bugcreekian" NALMA was proposed by Sloan (1987) and Archibald (1987a, 1987b) based on the time represented by the original Bug Creek faunas. A year later, I shared preliminary faunal data from McGuire Creek with J.D. Archibald which subsequently convinced him to abandon the "Bugcreekian" NALMA in favor of a less formalized mammalian zonation because of the difficulty of distinguishing earliest Puercan from "Bugcreekian" local faunas (see Archibald and Lofgren, 1990). Thus, the "Bugcreekian" NALMA was reduced in scale and became the Pu0 interval zone of the Puercan "Age," and McGuire Creek local faunas were referred to the Lancian and Puercan (Pu0 and Pu1 interval zones) NALMAs (Archibald and Lofgren, 1990) (Figure 11, Table 8). Additional faunal data from McGuire Creek (and sites nearby) further supports abandonment of a "Bugcreekian Age." These data and a discussion of the development of the "Bugcreekian" and the Pu0 interval zone (and subdivision into informal biochrons) are presented below.

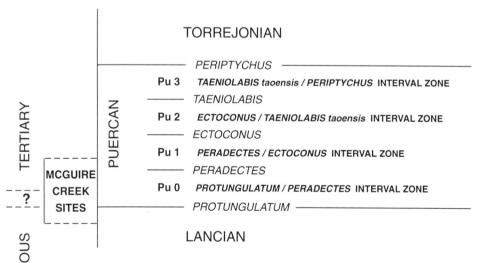

FIGURE 11. Relationship of McGuire Creek sites to previously proposed K-T mammalian zonations. Adapted from Archibald et al., 1987, and Archibald and Lofgren, 1990.

Table 8. Mammalian genera and dinosaur remains present at selected K-T local faunas from the western interior of North America[a][b]

[MCGUIRE CREEK LOCAL FAUNAS] spans columns LR–ZL.

| TAXA | FC | WK | BA | BW | HH | FR[c] | F1 | LF | LR | SL | BS | BG | SR | UU | ZL | HE | M |
|---|---|---|---|---|---|---|---|---|---|---|---|---|---|---|---|---|---|
| Dinosaurs | X | X | X | X | X | X | X | X | X | X | X | X | X | X |  | R |  |
| Paracimexomys | X | X |  |  |  |  |  |  |  |  |  |  |  |  |  |  |  |
| Alostera |  | X | X |  |  |  |  |  |  |  |  |  |  |  |  |  |  |
| Neoplagiaulax | X |  |  |  |  |  |  |  |  |  |  |  |  |  |  |  |  |
| Cimolodon | X | X | X | X |  |  | X | X |  |  |  | X |  |  |  |  |  |
| Cimolomys | X |  | X |  |  |  | X |  |  |  |  |  |  |  |  |  |  |
| Cimolestes | X | X | X |  |  |  | X | X |  |  |  |  |  |  |  |  |  |
| Essonodon | X | X | X |  |  |  |  |  | X |  |  |  |  |  |  |  |  |
| Batodon | X |  | X |  |  |  |  |  | X |  |  |  | X | X |  |  |  |
| Glasbius | X | X | X | X | X |  |  |  |  |  |  | X | X | X |  |  |  |
| Alphadon | X | X | X |  | X |  | X | X | X |  |  | X | X | X |  |  |  |
| Pediomys | X | X | X | X |  |  | X | X | X |  |  | X | X | X |  |  |  |
| Didelphodon | X | X | X | X |  |  |  | X | X |  |  | X | X | X |  |  |  |
| Gypsonictops | X | X | X | X | X |  | X | X | X |  |  | X | X | X |  |  |  |
| Meniscoessus | X | X | X | X | X |  | X | X | X | X | X | X | X | X |  |  |  |
| Cimexomys | X |  | X | X | X |  | X | X | X | X | X | X | X | X |  | X | X |
| Mesodma | X | X | X | X | X |  | X | X | X | X | X | X | X | X | X | X | X |
| Stygimys |  |  | X | X | X |  | X | X | X | X | X | X | X | X |  | X | X |
| Catopsalis |  |  | X | X | X |  | X | X |  | X | X | X | X | X |  | X | X |
| Procerberus |  |  | X | X | X |  | X | X | X | X | X | X | X | X | X | X | X |
| Protungulatum |  |  | X | X | X | X | X | X | X | X | X | X | X | X | X |  |  |
| Mimatuta |  |  | X | X | X | X |  | X | X | X | X | X | X | X | X | X | X |
| Oxyprimus |  |  | X |  | X | X |  | X | X | X |  | X | X | X |  | X | X |
| Ragnarok/Baioconodon |  |  | X | X |  |  | X | X | X | X | X | X | X | X |  | X | X |
| Peradectes |  |  |  |  |  |  |  |  |  |  |  | X | X | X | X | X | X |
| Maiorana, Earendil, Eoconodon, and Oxyacodon |  |  |  |  |  |  |  |  |  |  |  |  |  |  |  |  | X |

Abbreviation key and proposed biochronologic age (Puercan zonation after Archibald et al., 1987; Archibald and Lofgren, 1990). Lancian NALMA: FC=Flat Creek, WK=Wounded Knee. Puercan NALMA: *Protungulatum/Peradectes* INTERVAL ZONE Pu0 (formerly Bugcreekian): BA=Bug Creek Anthills, BW=Bug Creek West, HH=Harbicht Hill (Pu1?), FR=Ferguson Ranch (Pu1?), F1=Frenchman 1, LF=Long Fall, LR=Little Roundtop, SL=Second Level (Pu1?), BC=Black Spring Coulee (Pu1?). Puercan NALMA: *Peradectes/Ectoconus* INTERVAL ZONE Pu1: BG=Brown-Grey, SR=Shiprock, UU=Up-Up-the-Creek, HE=Hell's Hollow, ZL=Z Line, M=Mantua Lentil.

a. Location of local faunas: BA, BW, HH, FR, LR, SL, BS, BG, SR, UU, ZL, McCone County, Montana; FC, HE, Garfield County, Montana; M, Park County, Wyoming; WK, F1, LF, Saskatchewan, Canada.
b. Faunal data for McGuire Creek based on UCMP specimens (see Appendix 1). Other faunal data from: Archibald, 1982 (HE, FC, BA); Fox, 1987, 1989 (WK, F1, LF); Sloan and Van Valen, 1965 (BA, BW, HH); Van Valen, 1978 (BW, HH, M); Sloan et al., 1986 (FR); Jepsen, 1930, 1940 (M). Presence of *Glasbius*, *Alphadon*, and *Gypsonictops* at HH, and *Didelphodon* at BW, is based on study of UCMP specimens. Presence of *Mimatuta* and *Oxyprimus* at BA based on Luo, 1989, 1991.
c. Faunal data from Ferguson Ranch incomplete.
R= rare.

## THE "BUGCREEKIAN AGE"

The composition of their mammalian assemblages suggested that the Bug Creek faunas were younger than Lancian faunas (such as Ken's Saddle) within the Hell Creek Formation, but that they were deposited before the early Puercan (Sloan and Van Valen, 1965). Therefore, it was inferred that the Bug Creek faunas represented a previously unsampled faunal interval between the Lancian and Puercan "ages." Also, the Bug Creek faunas were placed in an ascending temporal sequence of Bug Creek Anthills-Bug Creek West-Harbicht Hill, partly on the basis of the gradual appearance of archaic ungulate mammal species (or "condylarths") (Bug Creek Anthills: *Protungulatum donnae*; Bug Creek West: *Protungulatum gorgun, Mimatuta morgoth*; Harbicht Hill: *Oxyprimus erikseni, Ragnarok "harbichti"*) (Sloan and Van Valen, 1965; Sloan et al., 1986).

However, following their original descriptions, the Bug Creek faunas were generally considered to be Lancian in age (Sloan, 1970, 1976; Russell, 1975; Fox, 1978; Archibald, 1982). Johnston (1980) reported vertebrate assemblages from Saskatchewan (Frenchman 1, Frenchman Formation; Long Fall, Ravenscrag Formation) similar in composition to the Bug Creek faunas. Later, these assemblages were also interpreted to be Lancian in age (Johnston and Fox, 1984; Fox 1987, 1989, 1990).

The first proposal that the Bug Creek faunas represented a new temporal unit following the Lancian was provided by Van Valen and Sloan (1977:42): "We will refer to the time from the deposition of Bug Creek Anthills to that of the Z coal bed as Bug Creek time". Later, reference was made to a "Bugcreekian Age" (Sloan, 1983; Rigby et al., 1986; and Sloan et al., 1986), but neither definition nor characterization were given.

Formal definition of a "Bugcreekian Age" was presented by Sloan (1987) and Archibald (1987a, 1987b). Sloan (1987) defined the "Bugcreekian Age (Stage)" as equivalent to the *Protungulatum* zone or the Bug Creek faunal sequence (Bug Creek Anthills-Bug Creek West-Harbicht Hill) and its equivalents (which included Frenchman 1). Archibald (1987a, 1987b) defined the "Bugcreekian" NALMA as the time between the successive first appearances of the "condylarth" *Protungulatum* and the marsupial *Peradectes*. However, R.C. Fox (1989) rejected the concept of a "Bugcreekian" NALMA, preferring to recognize the Frenchman 1 (and Long Fall) fauna as representing latest Lancian time.

Archibald et al. (1987) subdivided the Puercan NALMA into a series of interval zones defined by successive first appearances (or FADs: First Appearance Datums) of unrelated taxa. The initial interval zone (Pu1) of the Puercan was defined as the time between the first occurrence of the marsupial *Peradectes* and that of the "condylarth" *Ectoconus*. As defined, the Pu1 or *Peradectes/Ectoconus* interval zone included the Hell's Hollow Local Fauna (lower Tullock Formation of Garfield County, Montana; see Figure 1b).

## Pu0 AND Pu1 INTERVAL ZONE OF THE PUERCAN NALMA

The "Bugcreekian Age" of Archibald (1987a, 1987b) and Sloan (1987) was accepted as being a previously unsampled faunal interval between the Lancian and Puercan "ages"

Table 9. Occurrence of mammal genera and dinosaur remains in K-T biochronologic units

| | | | —————Pu0————— | | |
|---|---|---|---|---|---|
| | Lancian | bk1 | bk2 | bk3 | Pu1 |
| DINOSAUR REMAINS | X | X | X | X | X |
| **MULTITUBERCULATA** | | | | | |
| *Mesodma* | X | X | X | X | X |
| *Neoplagiaulax* | X | | | | X |
| *Cimolodon* | X | X | X | X | |
| *Meniscoessus* | X | X | X | X | X |
| *Cimolomys* | X | X | | X | |
| *Essonodon* | X | X | X | X | |
| *Cimexomys* | X | X | X | X | X |
| *Stygimys* | | X | X | X | X |
| *Catopsalis* | | X | X | X | X |
| *Acheronodon* | | | | | X |
| **MARSUPIALIA** | | | | | |
| *Alphadon* | X | X | ? | X | X |
| *Peradectes* | | | | | X |
| *Glasbius* | X | X | X | X | X |
| *Pediomys* | X | X | X | X | X |
| *Didelphodon* | X | X | X | X | X |
| **EUTHERIA** | | | | | |
| *Batodon* | X | X | X | X | X |
| *Gypsonictops* | X | X | X | X | X |
| *Cimolestes* | X | X | X | X | X |
| *Procerberus* | | X | X | X | X |
| **"CONDYLARTHRA"** | | | | | |
| *Protungulatum* | | X | X | X | X |
| *Ragnarok* | | | | X | X |
| *Baioconodon* | | | | | X |
| *Eoconodon* | | | | | X |
| *Oxyprimus* | | ? | | X | X |
| *Mimatuta* | | ? | X | X | X |
| *Earendil* | | | | | X |
| *Maiorana* | | | | | X |

*: Lancian genera whose presence in Pu0 and Pu1 sites is considered to be the result of reworking.
?: questionable record.

Table 10. Mammalian species and dinosaur remains present in local faunas at McGuire Creek[a]

| LOCAL FAUNA:<br>(NISP): | [—UPPER HELL CREEK FORMATION——] | | | | | | | | [TULLOCK FM] |
|---|---|---|---|---|---|---|---|---|---|
| | KM<br>(24) | LR[b]<br>(255) | SL[b]<br>(407) | BS<br>(27) | BG[b]<br>(832) | SR[b]<br>(408) | UU<br>(970) | ZL<br>(61) | JC<br>(8) |
| DINOSAUR REMAINS | X | X | X | X | X | X | X | | |
| **MULTITUBERCULATA** | | | | | | | | | |
| *Meniscoessus robustus* (43) | X | X | X | X | X | X | X | | |
| *Essonodon browni* (1) | | | X | | | | | | |
| *Cimolodon nitidus* (1) | | | | | X | | | | |
| *Mesodma* sp. (>2200) | | X | X | X | X | X | X | X | X |
| *Stygimys kuszmauli* (>450) | | X | X | X | X | X | X | | ? |
| *Cimexomys gratus* (158) | | X | X | X | X | X | X | | |
| *Cimexomys minor* (45) | | X | X | X | X | X | X | | |
| *Catopsalis alexanderi* (2) | | | | | | | X | | |
| *Catopsalis joyneri* (4) | | X | | | X | X | | | |
| *Catopsalis* sp. indet. (4) | | | | X | X | X | X | | |
| **MARSUPIALIA** | | | | | | | | | |
| *Pediomys hatcheri* (6) | | X | X | | X | | X | | |
| *Pediomys krejcii* (8) | | | X | | X | X | X | | |
| *Pediomys florencae* (8) | X | | X | | X | | X | | |
| *Pediomys elegans* (26) | | X | X | | X | X | X | | |
| *Pediomys* sp. indet. (4) | | | | | X | | X | | |
| *Alphadon "wilsoni"* (10) | | | X | | X | X | X | | |
| *Alphadon rhaister* (5) | | X | X | | X | X | X | | |
| *Alphadon* sp. indet. (9) | | X | | | X | | X | | |
| *Alphadon marshi* (1) | | | | | X | | | | |
| *Glasbius twitchelli* (17) | | | | | X | X | X | | |
| *Didelphodon vorax* (21) | X | X | X | | X | X | X | | |
| *Peradectes* cf *pusillus* (9) | | | | | X | X | X | X | |
| **EUTHERIA** | | | | | | | | | |
| *Gypsonictops illuminatus* (24) | X | X | X | | X | X | X | | |
| *Batodon tenuis* (4) | | X | | | | X | X | | |
| *Procerberus formicarum* (151) | | X | X | X | X | X | X | X | |
| *Procerberus* sp. (not *P. formicarum*) (1) | | | | | | | | | X |
| **"CONDYLARTHRA"** | | | | | | | | | |
| *Protungulatum donnae* (34) | | X | X | X | X | X | X | | |
| *Protungulatum gorgun* (27) | | X | X | X | X | X | X | | |
| *Mimatuta morgoth* (20) | | X | X | X | X | X | X | | |
| *Mimatuta minuial* (5) | | | X | | X | | X | | |
| *Mimatuta* sp. indet. (29) | | | | X | X | X | X | X | ? |
| *Ragnarok nordicum* (>50) | | X | X | X | X | X | X | | |
| *Oxyprimus erikseni* (59) | | X | X | | X | X | X | | |

a. NISP's given both for species (mammals) and local faunas.
b. Local fauna includes anthill sample; numbers of anthill specimens are: LR(155), SL(312), BG(366), and SR(308). See Appendix 2 for complete data for each local fauna and Table 8 for abbreviation key.
?. questionable record.

by Archibald and Lofgren (1990), but was rejected as the basis for proposing a new land-mammal age. Instead, the "Bugcreekian" was reduced in rank and defined as the initial interval zone of the Puercan or Pu0: *Protungulatum/Peradectes* interval zone. This change was necessitated by the discovery of Bug Creek assemblages at McGuire Creek which were transitional in composition between the "Bugcreekian Age" (now Pu0) and the earliest Puercan Pu1 interval zone. Prior to these discoveries, the youngest "Bugcreekian" fauna (Harbicht Hill) and the oldest Puercan local fauna (Hell's Hollow) were faunally distinct (approximately 50% overlap of species; see Archibald, 1982: table 59). The major faunal differences between the Hell's Hollow Local Fauna and the original Bug Creek sites were the disappearance of many Lancian taxa combined with the first appearance of *Peradectes* at Hell's Hollow (Archibald, 1982: table 59). Table 8 gives additional faunal data from the Bug Creek sites.

When considering the faunal data from McGuire Creek, it is difficult to draw a boundary between the "Bugcreekian" (or Pu0) and Pu1 interval zones, because they are similar in composition. Ignoring Lancian genera (which might be reworked), Pu0 faunas at LF, LR, SL, and BS are virtually identical to those at BG, SR, UU, and HE except that the latter localities yield *Peradectes*. Therefore, as defined, the Pu0 interval zone lacked "index" genera (Table 9) and contained only two possible "index" species, *Catopsalis joyneri* and *Purgatorius ceratops*. However, data from McGuire Creek indicate that *Peradectes* and *C. joyneri* occur in the Brown-Grey Local Fauna (Pu1) (Table 10), and the occurrence of *P. ceratops* at Harbicht Hill is suspect (Lillegraven and McKenna, 1986). Therefore, local faunas from the late "Bugcreekian" (Pu0) and early Pu1 interval zones can be nearly identical.

This uncertainty is compounded at McGuire Creek because *Peradectes* is rare. Therefore, the assignment of LR, SL, BS, BG, SR, and UU faunas to either the Pu0 or Pu1 interval zones might be influenced by sampling factors.

Finally, Pu0 and Pu1 sites are not well constrained biostratigraphically at McGuire Creek or elsewhere in the western interior. Documentation of the limited biostratigraphic control on Lancian, Bug Creek (Pu0: LR; Pu1: BG, SR, UU, SL?, BS?), and Puercan (Pu1: ZL, JC) local faunas at McGuire Creek was presented earlier. Therefore, while arguing that the Pu0 interval zone ("Bugcreekian") is a previously unsampled faunal interval, Archibald and Lofgren (1990) concluded that the potential difficulty in distinguishing Pu0 and earliest Pu1 faunas, stratigraphically (see Figure 12) and by faunal composition, is too weak a biostratigraphic basis from which to propose a new time unit on the scale of a NALMA.

The abandonment of the "Bugcreekian Age" does not mean that the Pu0 interval zone is not a distinct biochronologic unit; it is just a question of scale. The successive appearances of a number of new taxa (*Catopsalis, Stygimys, Procerberus, Protungulatum, Mimatuta, Oxyprimus, Ragnarok*) in the Pu0 interval zone and the presence of these taxa in local faunas that contain *Peradectes* (Pu1 interval zone) support a biochronologic sequence of Lancian-Pu0-Pu1 (Archibald and Lofgren, 1990). The problem is one of discriminating between Pu0 and Pu1 faunas, because all of these newly appearing genera persist into the Pu1 interval zone (see Table 9) and *Peradectes* can be rare (see Table 10).

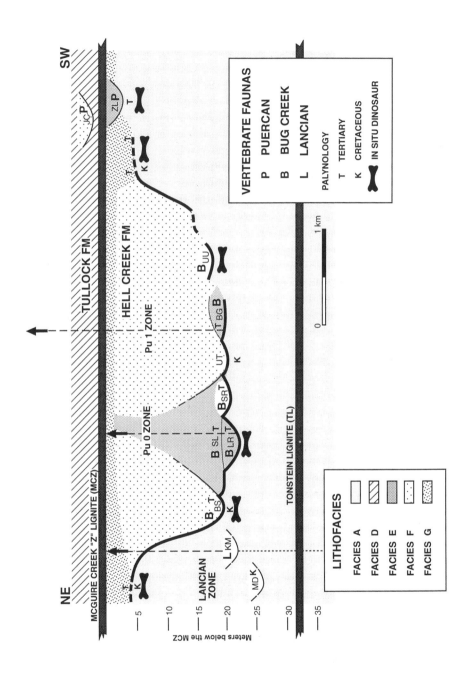

FIGURE 12. McGuire Creek biochronologic zonation based on biochronologic units proposed in Archibald et al., 1987, and Archibald and Lofgren, 1990. Channel fills or local faunas occur within the same stratigraphic interval. Note: All three biochronologic units occur within the same stratigraphic interval.

In addition to *Peradectes*, taxa whose first appearance are used to characterize the early Pu1 interval zone are *Ragnarok engdahli, Catopsalis alexanderi, Acheronodon garbani*, and *Mimatuta minuial*. However, *A. garbani* and *R. engdahli* are rare, which limits their use in regional correlations. The sample of *A. garbani* consists of the holotype, and *R. engdahli* is known only from the eight specimens described by Archibald (1982) from the Hell's Hollow Local Fauna (but see Rigby, 1989: figure 5).

*Catopsalis alexanderi* and *Mimatuta minuial* have a larger geographic distribution and sample size and therefore have greater utility for regional correlations. *C. alexanderi* is known from the Alexander locality, Denver Formation, Colorado, and Mantua Lentil, Polecat Bench Formation (Fort Union Formation), Wyoming, as well as the Hell Creek (this study) and Tullock formations, Montana (Middleton, 1982). *M. minuial* has a similar distribution, except that it is not known from the Denver Formation.

Recent work by Rigby (1989) indicates the occurrence of *M. minuial, R. engdahli*, and cf. *Peradectes* at the Doverzee locality (1 km north of the McGuire Creek study area). Doverzee and the local faunas from other localities under study by Rigby are probably referable to the Pu1 interval zone (FR, WT, DZ, VB, V1; Rigby, 1989: figure 5).

## INFORMAL BIOCHRONS, bk1-bk2-bk3, OF THE Pu0 INTERVAL ZONE

Archibald and Lofgren (1990) subdivided the Pu0 *Protungulatum/Peradectes* interval zone into three informal biochrons defined by first appearances of the "condylarths" *Protungulatum* (bk1), *Mimatuta* (bk2), and *Oxyprimus* (bk3). The type localities for the three informal biochrons are the original Bug Creek sites: Bug Creek Anthills (bk1), Bug Creek West (bk2), and Harbicht Hill (bk3). The three biochrons and correlations based on them were tentative because of the lack of stratigraphic control on the three local faunas, and the differences in faunal composition between the localities could be a factor of either inadequate sample size or lack of recent taxonomic reviews. Therefore, Archibald and Lofgren (1990) chose to propose the biochrons informally until additional faunal data from these sites is available. Analyses of these faunas is presently underway by J. K. Rigby, Jr., and it is hoped that the results of this work will be available soon.

With the uncertainties surrounding the type localities of the bk1-bk2-bk3 biochrons, it might be better to abandon hope of recognizing different faunal levels within the Pu0 interval zone. However, the Bug Creek Anthills Local Fauna is especially significant in this regard because it is older than other Pu0 sites in the upper Hell Creek Formation. Thousands of mammal teeth and jaws have been collected from Bug Creek Anthills and all systematists who have studied these mammalian samples agree that the larger-sized "condylarths" *Ragnarok* spp. or *Protungulatum gorgun* present at other Pu0 sites in the upper Hell Creek Formation are not present at Bug Creek Anthills (Tables 4,7,9). Given the huge sample size and similarity in facies between Bug Creek Anthills and other Pu0 sites in the upper Hell Creek Formation, the lack of *Ragnarok* spp. and *P. gorgun* at Bug Creek Anthills indicates that this local fauna is older.

However, analysis of faunal differences between other Pu0 sites is complicated by the stratigraphic, taxonomic, and sampling uncertainties discussed below.

## Lack of Stratigraphic Control

Stratigraphic relationships among the type localities for biochrons bk1 (Bug Creek Anthills), bk2 (Bug Creek West), and bk3 (Harbicht Hill) are uncertain. This is clearly the case with the Bug Creek Anthills locality, whose stratigraphic position has been debated at length (Smit and Van der Kaars, 1984; Fastovsky and Dott, 1986; Smit et al., 1987).

In spite of this, Sloan (1987:174) refers to the three original Bug Creek faunas as occurring within channel fills that are in superposition. In actuality, the Bug Creek West and Bug Creek Anthills local faunas occur within channel fills 2 km apart on opposite sides of the Bug Creek drainage and superposition of these channel fillings is not demonstrable (Fastovsky and Dott, 1986; Smit et al., 1987: figure 4; Rigby et al., 1987: figure 3). Also, the Harbicht Hill site is located over 20 km north of Bug Creek West and Bug Creek Anthills, and it is impossible to physically trace individual channel fills over 20 km in complex fluvial deposits such as those that comprise the upper Hell Creek Formation (Fastovsky and Dott, 1986; Fastovsky, 1987; this report).

As with the type localities, there is a general lack of superpositional or cross-cutting relationships among sites referable to the Pu0 interval zone. However, a series of localities in the Bug Creek drainage provides support for a bk1-bk2-bk3 succession. Tentative channel-fill relationships (Rigby et al., 1987) and preliminary faunal data (Rigby, 1989) indicate that the Big Bugger Channel yields *Oxyprimus* (bk3) and overlies the Bug Creek Anthills Channel, which yields only a single "condylarth," *Protungulatum* (bk1) (but see Luo, 1989, 1991). Also, the By George Channel yields *Oxyprimus* (bk3) and cuts the Scmenge Point Channel, which contains *Protungulatum* and *Mimatuta* but apparently not *Oxyprimus* (sensu bk2) (Rigby et al., 1987: figure 3; Rigby, 1989: figure 5). Therefore, channel fills with bk3 faunas overlie or cut channel fills with bk1 or bk2 faunas. The preliminary reports by Rigby et al. (1987) and Rigby (1989) are the only available biostratigraphic data which support a bk1-bk2-bk3 succession of informal biochrons.

## Sampling Uncertainties

Biochronologic zonations based on small faunal differences are particularly susceptible to sampling uncertainties. Successive first appearances of the "condylarths" *Protungulatum*, *Mimatuta*, and *Oxyprimus* define the respective biochrons bk1-bk2-bk3 of the Pu0 interval zone. McGuire Creek faunal data (mainly Pu1 sites) indicate that *Protungulatum*, *Mimatuta*, and *Oxyprimus* are not rare and are represented in roughly approximate numbers of specimens (Table 10). It is assumed that collection of a large sample would yield these "condylarth" genera if they were extant in the immediate area at the time the deposit was formed. The sample size from the Bug Creek Anthills locality is again critical in this regard, because the absence of certain taxa which are known to occur in other local faunas in the upper Hell Creek Formation strongly suggests that Bug Creek Anthills is older than other Pu0 sites.

However, the problem of insufficient sample size has been overlooked in previous biostratigraphic-biochronologic studies on the Bug Creek faunas. For example, a survey of the UCMP collection of mammals (n=300) from Harbicht Hill indicates that the marsupials *Alphadon* (UCMP 137313) and *Glasbius* (UCMP 137314) are present, as is the eutherian *Gypsonictops* (UCMP 137304) (Table 4). These taxa were thought to be extinct by the time the Harbicht Hill channel was filled (Sloan and Van Valen, 1965; Sloan, 1976). Similarly, the very small UCMP sample of mammals (n=10) from Bug Creek West yields an upper molar of *Didelphodon vorax* (Appendix 1). This is the first reported occurrence of *Didelphodon* from Bug Creek West. According to previous work, "*Didelphodon vorax*.....is quite unknown from any of the Paleocene localities in either Garfield or McCone counties" (sensu Bug Creek West and Harbicht Hill; Sloan and Rigby, 1986:1174). The former lack of records probably reflects small sample size because *D. vorax* and *Glasbius* are easily identified.

Last appearances of Lancian taxa such as *Didelphodon, Pediomys, Glasbius, Alphadon,* and others, and first appearances of certain "condylarths" from the Bug Creek West and Harbicht Hill sites have been accorded temporal significance (Sloan and Van Valen, 1965; Sloan, 1976; Van Valen and Sloan, 1977; Archibald, 1981, 1982, 1986, 1987c). The sizes of the samples from which these interpretations were made are not clear, although Sloan and Van Valen (1965) indicate that the minimum number of individuals (MNI) in their samples was 108 at Bug Creek West and 54 at Harbicht Hill. These samples are not large enough to include many of the rarer taxa. The sampling uncertainty is compounded by the fact that Bug Creek West and Harbicht Hill are channel-fill local faunas in which reworked fossils can occur. For example, samples from McGuire Creek indicate that virtually all Lancian taxa occur in Pu1 faunas (Table 4). However, many of these taxa are not found at Bug Creek West and Harbicht Hill which are Pu0 in age and presumably older than Pu1 sites. Is the presence of these taxa in McGuire Creek sites the result of reworking, or were Bug Creek West and Harbicht Hill not sufficiently sampled?

Based on the new faunal data presented above, the answer to the second question is yes. The reworking possibility was investigated earlier (see chapter above on Reworking of Fossils). In either case, it is apparent that the last appearance of Lancian taxa has limited value for biochronologic zonation. Also, the model of stepwise extinction of Lancian mammals (Van Valen and Sloan, 1977; Archibald, 1986, 1987c) within the Pu0 interval zone (Bug Creek Anthills-bk1, Bug Creek West-bk2, Harbicht Hill-bk3) is suspect because of their occurrence within the Pu1 interval zone (Tables 8, 9). Accordingly, Archibald and Lofgren (1990) abandoned last appearances of Lancian taxa as a basis for proposing or recognizing successive biochrons within the Pu0 interval zone.

Based on the examples discussed above, the available faunal lists of local faunas from the type localities of bk2 (Bug Creek West) and bk3 (Harbicht Hill) are far from complete, and what additional taxa these sites might eventually yield is unknown. The insufficient database from these localities raises doubt concerning the separability of the bk2 and bk3 biochrons.

Taxonomic Uncertainties

In 1965, announcement of the discovery and a brief description of mammalian components of the original Bug Creek faunas (Bug Creek Anthills, Bug Creek West, and Harbicht Hill) were published (Sloan and Van Valen, 1965). Other than "condylarths" from Bug Creek Anthills (Archibald, 1982; Luo, 1989, 1991), detailed systematic treatments of the mammalian faunas from the Bug Creek sites even now have yet to appear. Partly because of this lack of recent taxonomic review, proposal of a series of biochrons based on the mammal faunas from these localities was tentative (Archibald and Lofgren, 1990). Formal biochronologic zonations should be based on faunas that are well sampled, thoroughly described, and systematically analyzed. Therefore, mammals from the original Bug Creek sites require detailed description before informal biochrons (bk1-bk2-bk3) of the Pu0 interval zone can be either formally proposed or abandoned. Additional faunal data outlined below, based on UCMP collections support this contention.

*Bug Creek Anthills:* A survey of the multituberculates indicates that *Cimexomys gratus* (formerly *C. hausoi*) is represented by a few isolated teeth (see Appendix 1). Previously, only the smaller species of *Cimexomys*, *C. minor*, was thought to occur at Bug Creek Anthills (Sloan and Van Valen, 1965; Archibald, 1982).

Luo (1989, 1991) argued that *Mimatuta morgoth* and *Oxyprimus erikseni* are both present at Bug Creek Anthills, though rare. If his views are accepted, the bk2 and bk3 biochrons will require redefinition. In that case, the successive biochrons should be redefined as the *Protungulatum donnae/Protungulatum gorgun* bk1, the *Protungulatum gorgun/Ragnarok* bk2, and the *Ragnarok/Peradectes* bk3 (pending complete descriptions of the Bug Creek West and Harbicht Hill local faunas).

*Bug Creek West:* The UCMP sample of mammals is small (n=10) and adds no new data beyond the occurrence of *Didelphodon vorax* discussed above.

*Harbicht Hill:* The UCMP mammal sample (n=300) yields a few specimens of *Cimexomys gratus* and a single specimen of *Catopsalis* which is referable to *C. alexanderi* (Appendix 1). In the original description of the Harbicht Hill fauna, Sloan and Van Valen (1965) list only the smaller species of *Catopsalis*, *C. joyneri*. Either both species are present at Harbicht Hill or specimens identified as *C. joyneri* in 1965 are now referable to *C. alexanderi*, a taxon erected later by Middleton (1982). Analysis of samples of *Ragnarok nordicum* from Mantua Lentil and *R. harbichti* from Harbicht Hill indicates synonymy of these species, with *R. nordicum* having priority (Appendix 1).

In summary, the mammalian faunas from the type localities of the bk1-bk2-bk3 biochrons of the Pu0 interval zone are insufficiently described and sampled. Results of ongoing studies by J.K. Rigby, Jr., are required before these biochrons can be formally adopted or abandoned.

# CRETACEOUS-TERTIARY CORRELATIONS

## THE CRETACEOUS-TERTIARY BOUNDARY

Correlation of the Pu0 and Pu1 interval zones with the K-T boundary must be preceded by a means to recognize Cretaceous and Tertiary sediments within both marine and nonmarine sections that span the transition. The K-T boundary is usually placed between the Maastrichtian and Danian stages (but see Voight, 1981). However, the upper limit of the type section of the Maastrichtian Stage (ENCI Quarry, St. Pietersburg, South Holland) is an erosional surface (Felder et al., 1980), while the lower limit of the Danian Stage (Stevns Klint and Faxse, Denmark) disconformably overlies chalks correlated with the Maastrichtian (Berggren, 1964). Subsequently, several sections in Europe were recognized as lacking a major hiatus between the Maastrichtian and Danian stages or as being "relatively complete" (see Dingus, 1983: part 1, for discussion). Recently, the K-T section at El Kef, Tunisia, was proposed as the K-T boundary stratotype by the International Geological Congress in 1989. The marine K-T boundary at El Kef is defined by the first appearances of planktonic forminifera and calcareous nannoplankton (Keller, 1989; with recent modification of FADs and LADs of planktic foraminifera by MacLeod and Keller, 1991a, 1991b).

Biostratigraphic correlation between nonmarine and marine zonations is difficult because in few instances are fossiliferous marine and nonmarine interdigitations preserved and well exposed. This is the case with latest Cretaceous-earliest Tertiary marine (Fox Hills and Cannonball formations) and nonmarine (Hell Creek and Tullock-Ludlow formations) stratigraphic units of the western interior in the northern United States (Figure 2). The Hell Creek Formation is underlain by the Fox Hills Formation, whose mollusk zonation (Cobban, 1958) has been correlated with the European mollusk zonation of the Maastrichtian Stage (Jeletsky, 1960, 1962). The Tullock Formation is laterally equivalent to part of the Ludlow Formation, which overlies the Hell Creek Formation in the Dakotas (Carlson and Anderson, 1966; Moore, 1976; Belt et al., 1984). The Ludlow Formation interfingers with the Cannonball Formation (Jeletsky, 1962), which contains Danian planktonic foraminifera (Fox and Olsson, 1969). Therefore, the marine K-T boundary would fall somewhere within the upper Hell Creek or lower Tullock formations in eastern Montana.

A more precise biostratigraphic determination of the marine K-T boundary in this nonmarine sequence is possible using paleomagnetic and radiometric correlation methods. The marine K-T boundary at El Kef occurs within paleomagnetic anomaly C29R (Keller, 1989). The uppermost Hell Creek and lower Tullock formations in Garfield and McCone counties were deposited in a period of reversed polarity that is correlated with anomaly C29R (Archibald et al., 1982). In eastern Montana, this stratigraphic interval includes the last records of dinosaurs and most genera of Lancian mammals and the first occurrence of Pu0 and Pu1 mammal assemblages (Archibald, 1982; Archibald and Lofgren, 1990). Chron 29R is usually considered to span approximately 500,000 years (Harland et al., 1982), which makes marine-nonmarine correlation still relatively imprecise. Radiometric dating techniques provide little additional precision because when applied to latest Cretaceous samples standard errors can exceed 500,000 years.

Because of difficulties in identifying the stratigraphic level in nonmarine sequences that temporally corresponds to the marine K-T boundary, a number of operational definitions have been employed to identify the K-T boundary in terrestrial sediments in eastern Montana. These are: (1) the base of the first coal zone (the "Z" coal) above the highest occurrence of dinosaur remains (Brown, 1952); (2) the stratigraphically highest occurrence of dinosaur remains (Archibald, 1982); (3) palynofloral extinction or disappearance (Tschudy, 1970); and (4) iridium enrichment (Smit and Van der Kaars, 1984; Smit et al., 1987). These criteria for boundary recognition are reviewed below.

## "Z" Coal

This simple definition of using the first coal ("Z" coal) above the highest dinosaur was admittedly crude, but was proposed as a practical solution to end the prolonged debate during the first half of the twentieth century on where to place the K-T boundary in terrestrial sections in the western interior of North America (Brown, 1952). Subsequently, the "Z" coal was employed to approximate the K-T boundary in eastern Montana (Sloan and Van Valen, 1965; Van Valen and Sloan, 1977). However, the use of lithostratigraphic criteria to define a chronostratigraphic datum is both improper usage (Archibald, 1982; Fastovsky, 1987) and unrealistic. In a dynamic fluvial system, such as the one which deposited sediments spanning the K-T transition in eastern Montana, coal-swamp facies are not regionally persistent and coals are demonstrably discontinuous (Sholes and Cole, 1981; Archibald, 1982; Fastovsky, 1987).

This can be demonstrated in the McGuire Creek study area, where the TL and MCZ lignites are mappable for many kilometers but are not regional in extent. For example, the MCZ thins rapidly when traced to the northwest margin of the study area and grades laterally into black organic-rich mudstone (sections S, NN, OO, and HH, Plate 4), and correlations between sections OO and HH become highly questionable. Also, at Section OO (Plate 4), the "Z" coal of the Bug Creek drainage and the MCZ of the Black Spring Coulee drainage are apparently two different units (or part of a series of thin lignites that comprise the upper "Z" coal complex), and the "Z" coal is truncated by channeling. Therefore, both the concept of a regionally extensive "Z" coal bed and the use of the "Z" coal to define the K-T boundary should be discontinued.

## Highest Occurrence of Dinosaur Remains

This means of boundary recognition assumes that the passing of the last dinosaur signaled the close of the Cretaceous (Brown, 1952). Whether the timing of dinosaur extinction and the appearance of microfossils used to define the marine K-T boundary are synchronous is unknown. Also, if this means of boundary recognition were to be employed, then all local faunas containing unreworked dinosaur remains would be Cretaceous in age. However, the possibility that dinosaur remains in Pu0 and Pu1 assemblages from channel facies are reworked (and the difficulty in identifying reworked fossils) suggest that the highest occurrence of dinosaur remains may actually significantly postdate dinosaur extinction. Conversely, in any local stratigraphic section composed of fine-grained deposits of floodplain origin, the highest preserved occurrence of dinosaur remains may significantly predate the actual time that dinosaurs succumbed to extinction. Finally, to avoid circular reasoning, it is desirable to determine the timing of dinosaur extinction using independent criteria. Therefore, the use of the highest occurrence of dinosaur remains is neither demonstrably precise nor logically appropriate to employ for K-T correlations.

## Palynology

It has long been known that a palynological change could be recognized in local stratigraphic sequences in the western interior of North America that was roughly coincident with R. Brown's formula (1952) for the terrestrial K-T boundary (i.e., first coal above highest dinosaur) (Norton and Hall, 1969; Oltz, 1969; Leffingwell, 1970; Tschudy, 1970). More recently, discovery of an iridium enrichment at the marine K-T boundary in many sections worldwide (Alvarez et al., 1980; Orth et al., 1981, 1982; Alvarez et al., 1982, 1984; many others) has provided a means to globally correlate boundary events if the iridium enrichment represents the product of a single bolide impact (or another brief cataclysmic event) on earth. Recent K-T palynological research indicates that extinction of pollen species and an iridium enrichment are stratigraphically linked in many local sections throughout the western interior, from Saskatchewan and Alberta (Nichols et al., 1986; Jerzykiewicz and Sweet, 1986; Lerbekmo et al., 1987), Montana (Hotton, 1984, 1988; Bohor et al., 1984; Smit and van der Kaars, 1984; Smit et al., 1987), Wyoming (Bohor et al., 1987b), and Colorado-New Mexico (Orth et al., 1981; Tschudy et al., 1984). However, many other sections that are known to span the K-T transition in the western interior have been sampled for iridium but do not yield anomalous concentrations at the stratigraphic level that corresponds to the palynological K-T boundary.

The palynological K-T boundary in eastern Montana is recognized by the disappearance of Cretaceous indicator species, not by new appearances in the earliest Tertiary (Hotton, 1988). Absence of Cretaceous species can be employed to differentiate Cretaceous from Tertiary strata in local stratigraphic sections in eastern Montana (Sloan et al., 1986; Rigby et al., 1987; Smit et al., 1987; Hotton, 1988). Therefore, extinction of pollen species is an effective means to differentiate Cretaceous and Tertiary sediments, and was employed at McGuire Creek to determine the age of channel and floodplain facies yielding vertebrates.

## Iridium

Iridium enrichment (or concentrations of iridium above background levels) may provide the ideal criterion for correlating marine and nonmarine K-T boundary sections worldwide (Berry, 1984). High concentrations of iridium were first reported by Alvarez et al. (1980) at the marine K-T boundary (as recognized then) in Italy and Denmark, and later elsewhere worldwide (Alvarez et al., 1982, 1984). Alvarez et al. (1980) proposed that the iridium was too concentrated to have been terrestrially derived and therefore must be the record of a bolide impact. The effects of this impact on latest Cretaceous organisms were interpreted to be the causal factor in terminal Cretaceous extinctions, in both marine and terrestrial realms.

According to an impact scenario, extraterrestrially derived iridium would have been dispersed into the air after impact, and eventually incorporated into sedimentary basins within 100 years or less. Iridium-enriched sediment at K-T boundary sections worldwide would be the record of this virtually instantaneous event in earth history, and would thereby provide a basis for worldwide correlation (Berry, 1984). However, others argued that K-T iridium enrichment was deposited during a period of intense volcanism lasting 10,000 to 100,000 years (Officer and Drake, 1985; Officer et al., 1987; Crocket et al., 1988). Reports of high concentrations of iridium during recent volcanic emissions from Kilauea Volcano, Hawaii (Zoller et al., 1983), support the possibility that K-T iridium enrichment may be terrestrially derived. Also, the concentrating effects of micro-organismal activity may have had a role in the formation of iridium anomalies (Dyer et al., 1989).

Shocked quartz (Bohor et al., 1984; 1987a) and microspherules (Smit and Klaver, 1981; Montanari et al., 1983; Montanari, 1986) are often associated with iridium anomalies and may represent impact-derived products. However, shocked quartz may originate from intense volcanism (Carter et al., 1986; but see Alexopoulos et al., 1988), and microspherules may have a volcanic origin as well (Naslund et al., 1986). Also, microspherules from three marine K-T sites have been interpreted to represent diagenetic infillings of organic spheres, not impact or volcanism products (Hansen et al., 1986).

The source(s) of the iridium, shocked quartz, and microspherules near the K-T boundary no doubt will be debated for years. In this study, iridium enrichment is used as a correlation tool, whatever its source. Iridium enrichment is present at the proposed K-T boundary stratotype at El Kef, Tunisia (Kuslys and Krahenbulh, 1983). It also has been reported from many of the terrestrial stratigraphic sections known to span the K-T transition in the western interior of North America (Orth et al., 1981, 1982; Smit and Van der Kaars, 1984; Nichols et al., 1986; Lerbekmo and St. Louis, 1986; Lerbekmo et al., 1987; Bohor et al., 1987b). Other sections spanning the K-T transition, most notably those in eastern Garfield and western McCone counties in eastern Montana (where Bug Creek local faunas are located), have been sampled for iridium, but have not yielded a high level of iridium enrichment. In any case, the critical premise is that the marine and nonmarine iridium enrichments are records of the same discrete event, which allows precise correlation.

Several challenges to worldwide iridium anomaly correlations have been proposed. For example, one hypothesis suggests that K-T extinctions were caused by multiple impacts, perhaps comet showers (Hut et al., 1987). Also, the marine (worldwide distribution) and terrestrial (North American distribution) K-T iridium anomalies may have been deposited by different mechanisms (Schmitz, 1988). If the marine and non-marine anomalies were formed by different events, or if they were caused by multiple impacts, temporal correlation becomes suspect. However, iridium enrichment could still be an effective correlation tool for terrestrial K-T sections within North America.

Unfortunately, sections spanning the K-T transition in McCone County have been sampled for iridium enrichment without success (Smit et al., 1987). The nearest K-T iridium enrichment is 50 km to the west, at the Lerbekmo site in Garfield County. With the absence of records of iridium enrichment in McCone County, direct correlation of K-T sections to the marine K-T boundary cannot be accomplished. Pollen extinction then becomes the most precise correlation tool available for determining the local K-T boundary.

It has been implied that an iridium enrichment is present in the lower "Z" coal at Russell Basin, McCone County (Sloan et al., 1986; Rigby, 1989), a few kilometers north of the McGuire Creek study area. However, this iridium concentration is 40-80ppt, or only 2-3 times background (Fastovsky and Dott, 1986), and is similar to normal background levels such as those reported from Saskatchewan (5-60ppt, Nichols et al., 1986), North Dakota (25ppt, Johnson et al., 1989), and the Raton Basin in New Mexico (10-30ppt, Orth et al., 1981). A concentration of 40-80ppt is not anomalous, and reference to an iridium anomaly in McCone County should be discontinued.

## AGE OF THE Pu0 AND Pu1 INTERVAL ZONES

Pollen extinction is employed to differentiate Cretaceous and Tertiary sediment at McGuire Creek in the absence of iridium enrichment. Criteria for palynological differentiation of Cretaceous and Tertiary strata were developed from independently dated sequences of overbank deposits from both Garfield and McCone counties (Hotton, 1988). The data presented in an earlier chapter on palynological correlations show that all Pu0 and Pu1 sites at McGuire Creek yield Tertiary palynofloras (Table 3). Therefore, at McGuire Creek, the Puercan-Lancian and the palynological K-T boundaries coincide at the crude level of precision that is available to us: first appearance of *Protungulatum* = Pu0 interval zone; first record of loss of Cretaceous indicator palynomorphs = Tertiary.

Because all Pu0 and Pu1 vertebrate assemblages were collected from channel facies, and the palynological K-T boundary criteria were developed in overbank sequences, it was important to test whether a facies bias might exist. The composition of the flora contributing to the pollen rain along channel margins might differ significantly from that in overbank facies. The absence of Cretaceous indicator species in a Cretaceous channel might be due to these species' preference for overbank environmental settings. Therefore, a sample from the channel fill containing Matt's Dino Quarry (Section AA, Plate 3), which yielded associated hadrosaur skeletal remains, was analyzed for pollen. It yielded a Cretaceous palynoflora. This one sample suggests that palynological samples from channel facies are not biased by facies effects.

Erosion of Cretaceous strata and subsequent reworking of Cretaceous fossils into Paleocene channels occurred at McGuire Creek (Lofgren et al., 1990). A related issue is whether Cretaceous indicator palynomorphs were also reworked into these same channels. Paleocene channels yield a few occurrences of Cretaceous indicator species. These are usually less than 2% of total number counted, while the sample from Matt's Dino Quarry yielded 25% Cretaceous palynomorphs. For comparison, Cretaceous floodplain deposits yield 10-40% Cretaceous palynomorphs (pers. comm., C. Hotton, 1988). Therefore, it appears that if significant amounts of Cretaceous palynomorphs were eroded and redeposited, their numbers were diluted by large quantities of Tertiary pollen rain.

Reworking would explain the original conflict (now resolved) in age interpretation between the USGS palynologists and Carol Hotton concerning the Brown-Grey Channel. Thin siltstones in the Brown-Grey Channel were sampled and portions of a single sample were sent for analysis to two palynologists with experience in the K-T boundary problem. Rock sample 88DLL7-14-30 (original field no. 85H7-26-5, collected by J.H. Hutchison, 1985) was analyzed by palynologists at the USGS and subsequently assigned a Cretaceous age (pers. comm., Nichols and Wingate, March, 1988, USGS PL No. D6906). Sample 88DLL7-14-30 was also analyzed by Dr. Hotton, who concluded that it contains a Paleocene palynoflora (pers. comm., C. Hotton, December 1988).

This apparent conflict in age interpretation of 87DLL7-14-30 (=85H7-26-5) is a reflection of how the palynological boundary is defined. Earliest Paleocene palynofloras are depauperate and are recognized primarily on the absence or rarity of Cretaceous indicator species, not by new appearances in the earliest Paleocene (Hotton, 1988). Both analyses of the same rock sample indicate the presence of Cretaceous indicator species, but they are low in relative abundance (2.5-3.5% of the total sample), indicating a Tertiary age (pers. comm., C. Hotton, 1989). The USGS specialists made their Cretaceous interpretation based on the presence of Cretaceous indicator species. Therefore, the palynological boundary was recognized on different criteria by different palynologists: (1) the presence of Cretaceous indicator species no matter what their relative abundance; and (2) the relative abundance of Cretaceous indicator species.

In an attempt to resolve this disparity in approach, Dr. Nichols analyzed more samples of 88DLL7-14-30 and also 88DLL7-14-13, both from the Brown-Grey Channel. He concludes that because of the rare occurrence of characteristic Cretaceous palynomorphs in the samples, they still appear to be latest Cretaceous in age. However, the samples could be earliest Paleocene containing a few reworked Cretaceous palynomorphs. Identification of a species of pollen in the assemblage that may be restricted to the lowermost Paleocene in Wyoming and the Dakotas suggests this may be correct and earliest Paleocene is probably the most reasonable interpretation (pers. comm., D. Nichols, 1990). Therefore, Hotton and Nichols now agree that it is probable that the Brown-Grey Channel yields a Tertiary palynoflora with a low abundance of reworked Cretaceous palynomorphs (pers. comm., C. Hotton, 1989, and D. Nichols, 1990). This is similar to what may be occurring with the vertebrate assemblage from the same channel (i.e., reworking of dinosaurs and Lancian mammals).

Elsewhere in McCone County, Montana, where vertebrate sites yielding Pu0 and Pu1 have been sampled for pollen, most Pu0 and Pu1 sites yield Paleocene palynofloras (Sloan et al., 1986; Rigby et al., 1987; Newmann, 1988). However, Cretaceous palynofloras are claimed to be associated with Pu0 assemblages from channel fills at Bug Creek Anthills (Newmann, 1988) and Grenda's Horn/Doc's Folly (Rigby, 1989). If this is true, then the Pu0 interval zone would span the K-T boundary. However, detailed stratigraphic or palynological data were not given in these reports. The position within the channel fill of the Cretaceous pollen or from what lithology the pollen was collected (mud clast, or a siltstone lens, etc.), are not stated, nor are abundances/numbers of Cretaceous indicator species. Therefore, evaluation of claims of Cretaceous Pu0 assemblages (Newmann, 1988; Rigby, 1989) are not possible from the data given.

Similarly, two reportedly Pu0 assemblages (Frenchman 1 and Long Fall) from channel facies in Saskatchewan which yield dinosaurs and Lancian mammals are claimed to be Cretaceous in age, on the basis of relative stratigraphic position (Frenchman 1) or the presence or association with "unreworked" dinosaur and Lancian mammal remains (Frenchman 1 and Long Fall) (Johnston and Fox, 1984; Fox, 1987, 1989, 1990). However, Paleocene channels may contain reworked Cretaceous fossils (Lofgren et al., 1990). Also, palynological analysis of sediment within either channel fill has not been reported. Therefore, Pu0 assemblages from Frenchman 1 and Long Fall, as well as those from Bug Creek Anthills and Grenda's Horn/Doc's Folly, may be Cretaceous, but this remains to be demonstrated. If these Pu0 assemblages are Cretaceous, the Pu0 interval zone would span the K-T transition, and the Lancian-Puercan and Cretaceous-Tertiary boundaries would be diachronous.

# FAUNAL TURNOVER DURING THE
# K-T TRANSITION IN MONTANA

Bug Creek Assemblages (or Pu0 and Pu1 assemblages containing dinosaurs and Lancian mammals) are intermediate in composition when compared to unquestionably latest Cretaceous and earliest Paleocene vertebrate assemblages (Table 1). Estimates of vertebrate survival rates during the K-T transition in the western interior of North America can be significantly affected by two factors concerning these assemblages: (1) whether the Pu0 interval zone spans the K-T transition (or at least one Bug Creek assemblage is Cretaceous); and (2) whether the presence of dinosaurs and Lancian mammals in Pu0 and Pu1 assemblages is entirely the result of reworking.

The survival rate of non-dinosaurian lower vertebrates during the K-T transition is relatively high even if Bug Creek occurrences are omitted from the analysis. In eastern Montana, the survival rate of turtles spanning the Hell Creek-Tullock formations (and the K-T transition) is 84% at the generic level (Hutchison and Archibald, 1986). Similarly, the survival rate of species of non-dinosaurian lower vertebrates is at least 55% across the same stratigraphic interval (Bryant, 1989). If genera present elsewhere in the western interior during the Tertiary, but not yet found in the Tullock Formation, are included in the analysis, the survival rate of species could exceed 70% (Bryant, 1989).

In contrast, the survival rates of dinosaurian and mammalian vertebrates are significantly affected by the inclusion of Bug Creek occurrences in the analysis. Dinosaurs are represented by approximately 14-20 species in the Late Cretaceous Lancian "Age" (Sullivan, 1987; Archibald and Bryant, 1990). If dinosaurs did not persist into the Paleocene, then their survival rate is 0%. However, if bone fragments and isolated dinosaurian teeth present in channels that yield Paleocene palynofloras are accepted as evidence that dinosaurs survived into the Paleocene (an interpretation not favored here), the survival rate of dinosaur species would exceed 50% (13 of 20 species).

Mammalian survival rates are more complex because of the likelihood of reworking and the age uncertainty of the Bug Creek Anthills site. On the basis of data from eastern Montana (Table 11), if the Pu0 and Pu1 interval zones are entirely Paleocene and all the Cretaceous mammal species are reworked (i.e. those in question from Bug Creek assemblages), then the mammalian survival rate is 7% (2/28 species) or 13% (2/16 genera). In contrast, if all Cretaceous mammal species present in Pu0-Pu1 assemblages actually persisted into the Paleocene (were not reworked), then the survival rate

Table 11. Mammalian species from Cretaceous and early Paleocene local faunas in eastern Montana[a]

| | TYPICAL CRETACEOUS | If BA is CRETACEOUS | TYPICAL PALEOCENE |
|---|---|---|---|
| **MULTITUBERCULATA** | | | |
| *Meniscoessus robustus* | X | X | # |
| *Essonodon browni* | X | X | # |
| *Cimolodon nitidus* | X | X | # |
| *Cimolomys gracilis* | X | X | |
| *Paracimexomys priscus* | X | | |
| *Neoplagiaulax burgessi* | X | | X |
| *Cimexomys minor* | X | X | X |
| *Cimexomys gratus* | | * | X |
| *Mesodma thompsoni* | X | ? | |
| *Mesodma formosa* | X | ? | |
| *Mesodma hensleighi* | X | | |
| *Mesodma* sp. | | * | |
| *Mesodma garfieldensis* | | | X |
| *Stygimys kuszmauli* | | * | X |
| *Catopsalis joyneri* | | * | X |
| *Catopsalis alexanderi* | | | X |
| *Acheronodon garbani* | | | X |
| **MARSUPIALIA** | | | |
| *Pediomys hatcheri* | X | X | # |
| *Pediomys krejcii* | X | X | # |
| *Pediomys florencae* | X | X | # |
| *Pediomys elegans* | X | X | # |
| *Pediomys cooki* | cf | X | |
| *Alphadon "wilsoni"* | X | X | # |
| *Protalphadon lulli* | X | | |
| *Alphadon rhaister* | X | | # |
| *Alphadon marshi* | X | X | # |
| *Glasbius twitchelli* | X | X | # |
| *Didelphodon vorax* | X | X | # |
| *Peradectes pusillus* | | | cf |
| **EUTHERIA** | | | |
| *Gypsonictops illuminatus* | cf | X | # |
| *Batodon tenuis* | X | X | # |
| *Cimolestes magnus* | X | X | |
| *Cimolestes propalaeoryctes* | X | | |
| *Cimolestes stirtoni* | X | | |
| *Cimolestes incisus* | X | X | |
| *Cimolestes cerberoides* | cf | | |
| *Procerberus formicarum* | | * | X |
| *Protungulatum donnae* | | * | X |
| *Protungulatum gorgun* | | * | X |
| *Protungulatum mckeeveri* | | | X |
| *Mimatuta morgoth* | | * | X |
| *Mimatuta minuial* | | | X |
| *Ragnarok nordicum* | | | X |
| *Ragnarok engdahli* | | | X |
| *Oxyprimus erikseni* | | * | X |
| *Purgatorius ceratops* | | | X? |

Sources for taxonomic data are: Archibald, 1982; Sloan and Van Valen, 1965; Van Valen and Sloan, 1965; Novacek and Clemens, 1977; Luo, 1989, 1991; Archibald and Lofgren, 1990; and this study. The questionably Cretaceous local fauna is Bug Creek Anthills (see Newmann, 1988; Rigby, 1989).

a. Cretaceous and Tertiary are differentiated by palynology.

*: Questionable Cretaceous occurrences.

#. Paleocene Occurrences that may be due to reworking.

cf: Tentative referral.

---

rises to 57% (16/28 species) or 75% (12/16 genera). Finally, if the Bug Creek Anthills locality were actually Cretaceous and this faunal data were included in the analysis, the survival rate of mammals would climb to 65% (24/37 species) or 82% (18/22 genera).

Bug Creek assemblages are frequently cited as evidence that rates of vertebrate extinction/origination were gradual or stepwise during the K-T transition and are not compatible with catastrophic extinction scenarios (Van Valen and Sloan, 1977; Clemens and Archibald, 1980; Clemens et al., 1981; Archibald, 1981, 1982, 1984, 1987c; Clemens, 1982; Archibald and Clemens, 1984; Van Valen, 1984). However, estimates of vertebrate survival rates during the K-T transition in eastern Montana can be significantly affected by how the dinosaurian and mammalian components of Pu0 and Pu1 assemblages (or Bug Creek assemblages) are interpreted. If dinosaurs and Cretaceous mammals persisted into the Paleocene (not reworked) and/or Bug Creek Anthills is Cretaceous, the species survival rate of dinosaurs would exceed 50% and that of mammals 65%. If this is true, catastrophic mass-extinction scenarios involving terrestrial vertebrate species during the K-T transition (e.g., Smit and Van der Kaars, 1984; Smit et al., 1987) would be effectively falsified.

Data from McGuire Creek indicate that Bug Creek assemblages are Paleocene and contain reworked Lancian mammals and dinosaurs. If this interpretation is correct, the survival rate of mammal species is approximately 10% and of dinosaurs 0%, making catastrophic mass-extinction scenarios more plausible. However, such scenarios are still not compatible with the high specific survival rates (55-71%) of non-dinosaurian vertebrates (Bryant, 1989) or the entire vertebrate fauna (exceeding 50%) (Archibald and Bryant, 1990).

# CONCLUSIONS

The "Bug Creek Problem" refers to unresolved issues concerning the vertebrate biostratigraphy of the uppermost Hell Creek and lower Tullock formations of eastern Montana. This stratigraphic interval spans the K-T transition and yields three kinds of vertebrate assemblages: Lancian (Late Cretaceous), Bug Creek, and Puercan (Early Paleocene). Bug Creek assemblages are compositionally intermediate between Lancian and Puercan faunas, in that they contain Lancian mammals and dinosaurs along with mammals whose descendants are characteristic of the Puercan; thus they may be either Cretaceous, Paleocene, or both in age. Detailed systematic and biostratigraphic treatments of Bug Creek assemblages in concert with geologic analyses of the stratigraphic interval in which they occur were needed because these assemblages are central to debates concerning patterns of faunal turnover of the terrestrial biota during the Cretaceous-Tertiary transition.

Uncertainties concerning Bug Creek assemblages can be divided into four main issues, which are: (1) Are Bug Creek assemblages temporally intermediate between Lancian (Late Cretaceous) and Puercan (Early Paleocene) assemblages, on the basis of biostratigraphic and/or biochronologic criteria? (2) Can Bug Creek assemblages themselves be placed in a temporal sequence that reflects faunal differences between individual local faunas? (3) Do Bug Creek assemblages represent the remains of vertebrates that lived contemporaneously, or were Cretaceous (Lancian) fossils eroded and redeposited with Paleocene (Puercan) vertebrate remains? (4) What are the geochronologic age(s) of Bug Creek assemblages?

Lithofacies and biostratigraphic analyses of sections spanning the K-T transition at McGuire Creek, McCone County, Montana, were used to address the "Bug Creek Problem." Lithofacies analysis of the upper Hell Creek and lower Tullock formations indicates that Bug Creek assemblages are exclusively associated with lag deposits of large channel facies that are deeply entrenched into floodplains deposits which yield in situ dinosaur remains. A locally traceable erosion surface was created by these channeling events. Because of the enormous amount of older sediment eroded during channel entrenchment, the potential is very high that Bug Creek assemblages contain reworked fossils.

An indication that the dinosaurian component of Bug Creek assemblages is reworked comes from the lateral tracing of channel facies into floodplains containing in situ dinosaur remains. In two instances where channel fills containing dinosaur remains can be physically traced into floodplain deposits, dinosaurs are absent in contemporaneous floodplain deposits.

At McGuire Creek, palynology is useful for the correlation of isolated channel fills containing Bug Creek assemblages with floodplain deposits yielding or lacking dinosaur remains. Where data are available, channel fills containing Bug Creek assemblages all yield Paleocene palynofloras, and floodplain deposits that contain the highest in situ dinosaur remains yield Cretaceous palynofloras. Also, the lowest Paleocene floodplain deposits identified palynologically lack dinosaurs. Therefore, both palynological and geological correlations indicate that channel fills containing Bug Creek assemblages are temporally equivalent to floodplain deposits which overlie those containing in situ dinosaurs. Also, the erosion surface developed between floodplain facies containing in situ dinosaurs and channel fills yielding Bug Creek assemblages is the local demonstration of the palynological Cretaceous-Tertiary boundary.

Because channel facies that yield Bug Creek assemblages are deeply incised into Lancian strata, considerable reworking of dinosaur and Lancian mammal remains must have occurred. However, fossils that were reworked can be identified only in exceptional instances, such as Black Spring Coulee, where many large dinosaur bones were reworked into a Paleocene channel fill (Lofgren et al., 1990). Because of the great potential for reworking, fossils representing all suspect Bug Creek taxa (dinosaurs and most species of Lancian mammals) should be assumed to be reworked unless at least one of these criteria are met: (1) articulated skeletal remains occur within a channel fill; (2) specimens are recovered from Paleocene floodplain deposits; (3) a channel that did not scour Lancian or Cretaceous sediments yields abundant specimens of suspect taxa; (4) specimens are demonstrably reworked from Puercan or Paleocene strata.

Development of a local biostratigraphic zonation for the uppermost Hell Creek Formation that subdivides strata in accordance with the presence of Lancian, Bug Creek, and Puercan assemblages is difficult because they occur in channel facies within a complexly channeled stratigraphic interval, and only in rare instances can channels be temporally ordered based on superposition or crosscutting relationships.

On the basis of previously proposed biochronologic zonations (Archibald et al., 1987; Archibald and Lofgren, 1990), McGuire Creek vertebrate assemblages are referred to the Lancian NALMA and Pu0 and Pu1 interval zones of the Puercan NALMA. Although most Pu0 and Pu1 sites are not well constrained stratigraphically, the successive appearances of new taxa (*Catopsalis, Stygimys, Procerberus, Protungulatum, Mimatuta, Oxyprimus, Ragnarok*) in the Pu0 interval zone and the presence of these taxa into local faunas that contain *Peradectes* (Pu1 interval zone), support a biochronologic sequence of Lancian-Pu0-Pu1 (Archibald and Lofgren, 1990).

The informal division of the Pu0 interval zones into three informal biochrons (bk1-bk2-bk3; Archibald and Lofgren, 1990) was tentative because the mammalian faunas from the type localities of these biochrons (Bug Creek Anthills-bk1, Bug Creek West-bk2, Harbicht Hill-bk3) either require additional sampling or more recent systematic treatments. Analyses of these local faunas are presently underway, and results of these studies are required before these biochrons can be formally adopted or abandoned.

Extinction of Cretaceous palynomorph species has been employed to recognize earliest Tertiary sediments in eastern Montana (Hotton, 1988). Therefore, palynological criteria were employed to differentiate Cretaceous and Paleocene channel fills at McGuire Creek. Where palynological data are available, all channel fills that yield Puercan (Pu0 and Pu1) assemblages at McGuire Creek are Paleocene. The Brown-Grey Channel was originally considered to contain a palynoflora that could be interpreted, from independent palynological analyses of a single rock sample, as either Cretaceous or Paleocene. The conflict stemmed from different definitions of the palynological boundary: (1) a very low relative abundance of Cretaceous indicator species (Paleocene); or (2) the presence of such species no matter how rare (Cretaceous). While consensus on a Paleocene age assignment was reached, the original conflict suggests that the palynological K-T boundary can be recognized by multiple criteria.

The age of Bug Creek assemblages (or Pu0-Pu1 assemblages with dinosaurs and certain Lancian mammals), and whether or not these assemblages contain reworked dinosaur and Lancian mammal remains, are factors that can significantly affect estimates of the survival rates of these vertebrates during the K-T transition in eastern Montana. Survival rates of dinosaurs and mammals can be interpreted to be compatible with catastrophic extinction scenarios if Bug Creek assemblages are exclusively Paleocene and contain reworked Lancian mammal and dinosaur remains because the survival rate of mammals would be approximately 7% (2/28 species), and of dinosaurs 0%. These interpretations are supported by analyses at McGuire Creek, where all Bug Creek assemblages are palynologically Paleocene and appear to contain Cretaceous vertebrates (those with suspect Paleocene records) because of reworking.

# Appendix 1
# Mammalian Systematics

Brief descriptions of mammalian components of the original Bug Creek assemblages were given by Sloan and Van Valen (1965), Van Valen and Sloan (1965), and Van Valen (1978). Except for the few studies noted here, detailed systematic treatments have yet to appear. Lillegraven (1969) provided additional description of *Procerberus*. Archibald (1982) described some of the mammals found in the Bug Creek assemblages, but the focus of his systematic work was on mammals from Garfield County. Novacek and Clemens (1977) studied the UCMP sample of *Mesodma* from the Bug Creek Anthills locality and argued that only a single species was discernable (contra Sloan and Van Valen, 1965). Luo (1989, 1991) studied the "condylarth" component of the UCMP sample from Bug Creek Anthills and argued that three species are present rather than the single species, *Protungulatum donnae*, described by Sloan and Van Valen (1965). Lupton et al. (1980) briefly described two partial dentaries of *Protungulatum gorgun* from Chris' Bonebed, a Milwaukee Public Museum locality in the upper Hell Creek Formation near Harbicht Hill (Figure 1c).

The original Bug Creek faunas (Bug Creek Anthills, Bug Creek West, Harbicht Hill) formed the basis for the "Bugcreekian" NALMA (Sloan, 1987; Archibald, 1987b), but are now recognized as comprising the initial interval zone (Pu0) of the Puercan Age (Archibald and Lofgren, 1990).

Lancian mammals whose presence in Puercan faunas (Pu0 or Pu1 interval zones) in eastern Montana may be entirely caused by reworking are: *Meniscoessus robustus, Essonodon browni, Cimolodon nitidus, Pediomys hatcheri, P. krejcii, P. florencae, P. elegans, Pediomys* sp. indet., *Alphadon "wilsoni," A. rhaister, A. marshi, Alphadon* sp. indet., *Glasbius twitchelli, Didelphodon vorax, Gypsonictops illuminatus*, and *Batodon tenuis*. These questionable occurrences in Puercan faunas are denoted by an asterisk (*) in the listed distribution of each species. *Cimexomys minor* and *Neoplagiaulax burgessi* are Lancian mammals whose present in Puercan faunas apparently is not the result of reworking from Cretaceous sediments.

Lithostratigraphic units that yield both Puercan and Lancian mammalian local faunas are the Hell Creek Formation in McCone County, Montana, and the Frenchman Formation in Saskatchewan, Canada.

Class MAMMALIA Linnaeus, 1758
Subclass ALLOTHERIA Marsh, 1880
Order MULTITUBERCULATA Cope, 1884
Suborder PTILODONTOIDEA (Gregory and Simpson, 1926) Sloan and Van Valen, 1965
Family NEOPLAGIAULACIDAE Ameghino, 1890
*Mesodma* Jepsen, 1940
*Mesodma* sp.

*Comments*: A large number (n>2200) of isolated and associated teeth from fragmentary dentaries and maxillas referable to *Mesodma* were recovered from localities in the upper Hell Creek Formation at McGuire Creek sampled by screenwashing techniques. Distinction of species of *Mesodma* is a difficult taxonomic problem (Sloan and Van Valen, 1965; Novacek and Clemens, 1977; Archibald, 1982). Because of the large number of fossils and the difficulties inherent in species identification, a thorough study of the sample of *Mesodma* from McGuire Creek is beyond the scope of this study. Although it is uncertain if one or more species are present, all specimens are referred to *Mesodma* sp. at this time.

Family CIMOLODONTIDAE Marsh, 1889a
*Cimolodon* Marsh, 1889a
*Cimolodon nitidus* Marsh, 1889a

*Cimolodon nitidus* Marsh, 1889a, p. 84 (see Clemens, 1964, p. 56, for synonymies).
*Holotype.* YPM 11776, left /M1 (Marsh, 1889a, pl. II, figs. 5-8).
*Type locality.* Mammal locality no. 1 of Lull (1915), UCMP loc. V5003, Lance Formation, Wyoming.
*Referred specimen.* 1 M1/, UCMP 133406, from loc. V87072.
*Locality.* UCMP loc. V87072.
*Distribution.* St. Mary River Formation, Alberta (Edmontonian); Lance Formation, Wyoming, and Scollard Formation, Alberta (both Lancian); Frenchman Formation, Saskatchewan, and Hell Creek Formation, Montana (both Lancian and *Puercan); Ravenscrag Formation, Saskatchewan (*Puercan).
*Revised diagnosis.* Clemens, 1964, p. 56.
*Description.* The single tooth referable to *Cimolodon nitidus* is an M1/. The cusp formula (5:7:5) and size (L: 4.51, W: 2.57) of this specimen agree with Clemens' (1964) description of the species.

Suborder TAENIOLABIDOIDEA (Granger and Simpson, 1929)
Sloan and Van Valen, 1965
Family EUCOSMODONTIDAE (Jepsen, 1940)
Sloan and Van Valen, 1965
Subfamily EUCOSMODONTINAE Jepsen, 1940
*Stygimys* Sloan and Van Valen, 1965
*Stygimys kuszmauli* Sloan and Van Valen, 1965
Tables 12-13; Figures 13-19

*Eucosmodon gratus* Jepsen, 1930, p. 499-500.

*Stygimys gratus* Sloan and Van Valen, 1965, p. 224.

*Holotype.* UMVP 1478, left lower jaw fragment with I and alveoli for /P4 and /M1 (Sloan and Van Valen, 1965, part of fig. 4).

*Type locality.* Bug Creek Anthills, Hell Creek Formation, Montana.

*Referred specimens.* Three /P4's, 4 /M1's, 2 M1/'s, 3 /M2's, 7 M2/'s from loc. V87035. Six P4/'s, 4 /P4's, 8 /M1's, 9 M1/'s, 2 /M2's, 9 M2/'s from loc. V87037. Five P4/'s, 12 /M1's, 13 M1/'s, 6 /M2's, 11 M2/'s, from loc. V87038. Two P4/'s, 1 /P4, 1 /M1, 9 M1/'s, 4 /M2's, 4 M2/'s from loc. V87051. Three P4/'s, 5 /P4's, 5 /M1's, 4 M1/'s, 10 /M2's, 6 M2/'s from loc. V87071. Eight P4/'s, 10 /P4's, 19 /M1's, 18 M1/'s, 14 /M2's, 22 M2/'s from loc. V87072. Six P4/'s, 3 /P4's, 7 /M1's, 4 M1/'s, 9 /M2's, 14 M2/'s from loc. V87074. One P4/, 4 /P4's, 1 /M1, 5 M1/'s, 2 M2/'s from loc. V87077. Three P4/'s, 2 /P4's, 12 /M1's, 6 M1/'s, 8 /M2's, 6 M2/'s from loc. V87098. Two P4/'s, 8 /P4's, 14 /M1's, 10 M1/'s, 9 /M2's, 13 M2/'s from loc. V87151. Twenty five other McGuire Creek localities yielded 1-6 isolated teeth per site.

*Localities.* V86031, V87028, V87030, V87034, V87035, V87036, V87037, V87038, V87040, V87049, V87051, V87052, V87070, V87071, V87072, V87073, V87074, V87077, V87078, V87084, V87086, V87098, V87101, V87108, V87109, V87114, V87115, V87124, V87151, V87152, V87153, V88038, V88039, and V88041.

*Distribution.* Upper Hell Creek and lower Tullock formations, Montana, and Polecat Bench Formation (Fort Union Formation), Wyoming (all Puercan).

*Discussion and Description.* When first proposed, *Stygimys* included three species formerly referred to *Eucosmodon* (*E. gratus, E. jepseni, E. teilhardi*) and a new species, *Stygimys kuszmauli*, which was designated as the genotype (Sloan and Van Valen, 1965). *Stygimys kuszmauli* and *S. gratus* were reported from the Bug Creek sequence of channel deposits (Bug Creek Anthills, Bug Creek West, Harbicht Hill) in the upper Hell Creek Formation of Montana (Sloan and Van Valen, 1965). Sloan and Van Valen (1965) noted the presence of *S. gratus* at Harbicht Hill, which was previously known only from Mantua Lentil, Polecat Bench Formation, Wyoming (Jepsen, 1930). The new, smaller species of the genus, *S. kuszmauli*, was listed as occurring at Bug Creek Anthills (type locality) and Bug Creek West (Sloan and Van Valen, 1965).

The original diagnosis of *S. kuszmauli* is short and only refers to size differences between it and *S. gratus*: "This is the smallest species of the genus. Length /P4, 4.6 +/- .3mm (standard deviation of sample); 11 serrations. The specimens of *Stygimys* from Harbicht Hill are larger than those from Bug Creek Anthills and can be referred to *Stygimys gratus*" (Sloan and Van Valen, 1965, p. 224). The /P4's of *S. gratus* from Mantua Lentil have 11 serrations and lengths of 4.8mm and 4.9mm (Jepsen, 1940). These lengths fall within one standard deviation of the mean length of /P4's of *S. kuszmauli* from Bug Creek Anthills (Table 12). Therefore, the original diagnosis of *S. kuszmauli* does not distinguish it from *S. gratus*.

To resolve this problem, the samples of *Stygimys* in the UCMP collections from Bug Creek Anthills (V65127 and V70201) and Harbicht Hill (V71203) were compared to the small (n=6) sample of *Stygimys gratus* from Mantua Lentil described by Jepsen (1930; 1940). This analysis indicates that the holotype of *S. gratus*, PU 13373 a dentary

Table 12. Measurements of isolated molars and premolars of *Stygimys kuszmauli* from Bug Creek Anthills (V65127 and V70201, a sublocality within V65127), Harbicht Hill (V71203), and Mantua Lentil

|  | V65127-V70102 | V71203 | Mantua Lentil |
|---|---|---|---|
| **/P4 Length** | | | |
| Number | 37 | 8 | 2 |
| Observed Range | 3.82-5.46 | 4.30-5.33 | 4.65-4.68 |
| Mean | 4.67 | 5.10 | 4.67 |
| **/P4 Width** | | | |
| Number | 34 | 7 | 1 |
| Observed Range | 1.46-2.13 | 1.61-2.10 | 1.72 |
| Mean | 1.80 | 2.00 | 1.72 |
| **/M1 Length** | | | |
| Number | 42 | 12 | 1 |
| Observed Range | 3.62-4.79 | 3.83-4.43 | 4.32 |
| Mean | 4.19 | 4.13 | 4.32 |
| **/M1 Width** | | | |
| Number | 42 | 12 | 1 |
| Observed Range | 1.61-2.06 | 1.69-1.97 | 1.87 |
| Mean | 1.83 | 1.80 | 1.87 |
| **/M2 Length** | | | |
| Number | 46 | 10 | — |
| Observed Range | 2.38-2.96 | 2.49-3.17 | — |
| Mean | 2.70 | 2.80 | — |
| **/M2 Width** | | | |
| Number | 46 | 10 | — |
| Observed Range | 1.79-2.22 | 1.92-2.27 | — |
| Mean | 2.02 | 2.09 | — |
| **P4/ Length** | | | |
| Number | 37 | 3 | — |
| Observed Range | 2.91-3.92 | 3.14-3.49 | — |
| Mean | 3.34 | 3.32 | — |
| **P4/ Width** | | | |
| Number | 37 | 3 | — |
| Observed Range | 1.53-2.32 | 1.73-1.84 | — |
| Mean | 1.79 | 1.79 | — |
| **M1/ Length** | | | |
| Number | 31 | 2 | 2 |
| Observed Range | 4.31-5.35 | 4.74-4.78 | 5.23-5.25 |
| Mean | 4.83 | 4.76 | 5.24 |
| **M1/ Width** | | | |
| Number | 31 | 2 | 2 |
| Observed Range | 2.09-2.74 | 2.23-2.29 | 2.55-2.57 |
| Mean | 2.44 | 2.26 | 2.56 |
| **M2/ Length** | | | |
| Number | 28 | 10 | — |
| Observed Range | 2.33-2.88 | 2.65-3.05 | — |
| Mean | 2.67 | 2.79 | — |
| **M2/ Width** | | | |
| Number | 27 | 10 | — |
| Observed Range | 2.21-2.72 | 2.39-2.68 | — |
| Mean | 2.46 | 2.56 | — |

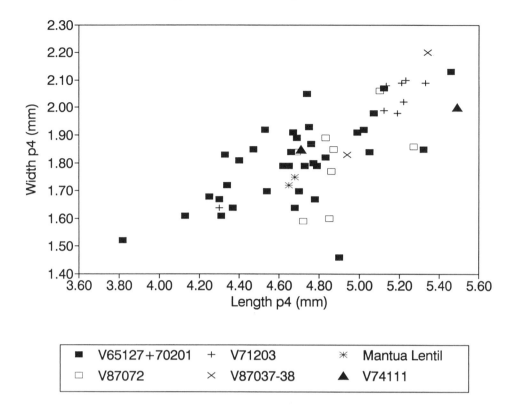

FIGURE 13. Bivariate plot (length vs. width) of /P4's of *Stygimys* from Bug Creek Anthills (V65127 and V70201), Harbicht Hill (V71203), Tedrow Quarry D (V87072), and Up-Up-the-Creek 2 and 3 (V87037-38) (all from the Hell Creek Formation, McCone County, Montana), Worm Coulee 1 (V74111; from the Tullock Formation, Garfield County, Montana), and Mantua Lentil (from the Polecat Bench Formation, Wyoming). One of the two /P4's from Mantua Lentil (PU 14496) is encased in sediment, and only length could be measured; its width was estimated at 1.75 mm, based on the similar length of PU 14419, which has a width of 1.72 mm.

fragment with /M2, represents a different taxon than the remainder of its hypodigm. The holotype (PU 13373) is much smaller than the two other dentary fragments from Mantua Lentil referred to *S. gratus*. Also, although the /M2 from the holotype is heavily worn, its morphology differs significantly from /M2's of *S. kuszmauli* from Bug Creek Anthills. Comparison of the holotype of *S. gratus* with that of *Cimexomys hausoi* from the Tullock formation of Garfield County, Montana, indicates that these two specimens represent the same taxon. Therefore, PU 13373 becomes the namebearer for that species, now recognized as *Cimexomys gratus* (further discussion in section on *C. gratus*).

Because the holotype of "*Stygimys gratus*" becomes the namebearer of *Cimexomys gratus*, the remainder of the Mantua Lentil sample of *Stygimys* is without species referral. Since the original diagnosis of *Stygimys kuszmauli* did not distinguish it from the Mantua Lentil sample of *Stygimys*, comparison of samples from the two type localities, Mantua

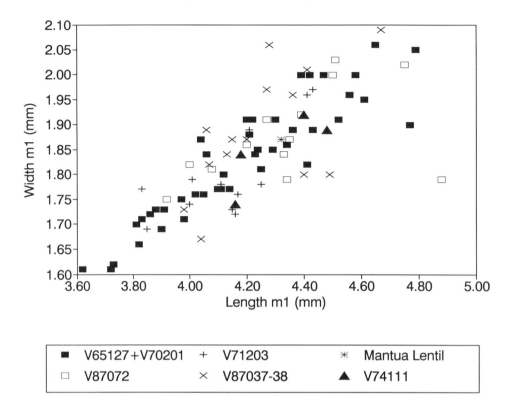

FIGURE 14. Bivariate plot (length vs. width) of /M1's of *Stygimys* from various localities (see Figure 13).

Lentil and Bug Creek Anthills, was required. The UCMP sample of *Stygimys* from Harbicht Hill was included in the analysis because when *Stygimys* was described initially, the Harbicht Hill sample was referred to "*Stygimys gratus*" (Sloan and Van Valen, 1965).

/P4: Both /P4's (PU 14496, PU 14419) from Mantua Lentil have 11 serrations. All /P4's from Harbicht Hill have 11 serrations. Three of 36 /P4's from Bug Creek Anthills have 10 serrations, the rest have 11. Based on size, /P4's from Mantua Lentil and Bug Creek Anthills are, on average, nearly identical (Table 12). A bivariate graph of length vs. width indicates that the two /P4's from Mantua Lentil fall within a cluster of specimens of *S. kuszmauli* from Bug Creek Anthills (Figure 13). The /P4 samples from Harbicht Hill and Bug Creek Anthills have similar ranges in size, but the mean /P4 length of specimens from Harbicht Hill is much larger than that from Bug Creek Anthills (Table 12). However, the difference in mean lengths of /P4's from Harbicht Hill (V71203) and Bug Creek Anthills (V65127 only) was tested statistically, and was not significant (T test: T= -1.23; df 16+ 2.12).

/M1: The single /M1 (PU 14417) from Mantua Lentil has a cusp formula of 6:5. Fifteen percent (7 of 45) of /M1's from Bug Creek Anthills have 6 cusps on the external row, the remainder have seven. All /M1's from Bug Creek Anthills have 5 cusps on the

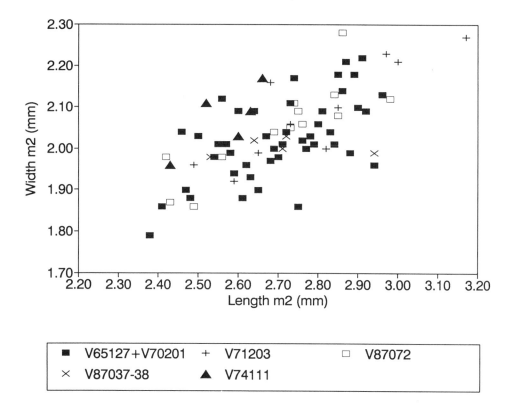

FIGURE 15. Bivariate plot (length vs. width) of /M2's of *Stygimys* from various localities (see Figure 13).

internal row. All /M1's from Harbicht Hill (n=10) have 5 cusps on the internal row, and all but one have 7 cusps on the external row; the one exception has 6. Based on size, all three samples are indistinguishable (Table 12). Plotted graphically by length vs. width, /M1's from Harbicht Hill and Mantua Lentil compare closely with those from Bug Creek Anthills (Figure 14).

/M2: The Mantua Lentil sample yields an /M2 (PU 13373) but it is referable to *Cimexomys* (this report), not *Stygimys* (contra Jepsen, 1930; sensu Sloan and Van Valen, 1965). Harbicht Hill /M2's (n=10) have a cusp formula of 4:2, as do Bug Creek Anthills /M2's (n=46) except for three, which have 5 cusps on their external rows. Based on size, /M2's from Harbicht Hill are, on average, slightly larger than those from Bug Creek Anthills (Table 12). However, when the two samples are plotted graphically, the small differences in mean length and width are apparently insignificant (Figure 15).

P4/: The Mantua Lentil sample does not contain P4/'s. Harbicht Hill P4/'s (n=3) have a cusp formula of 2-3:7-8:1, and those from Bug Creek Anthills (n=38) 2-3:7-10:1. Based on size, the samples are nearly identical (Table 12). A bivariate length vs. width plot indicates that the Harbicht Hill sample falls completely within the sample from Bug Creek Anthills (Figure 16).

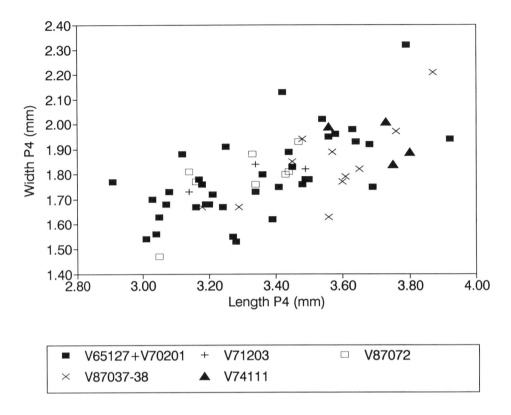

FIGURE 16. Bivariate plot (length vs. width) of P4/'s of *Stygimys* from various localities (see Figure 13).

M1/: The two M1/'s (PU 14420A, PU 14420B) from Mantua Lentil are significantly larger than the mean of those from Bug Creek Anthills and Harbicht Hill (Table 12) (PU 14420 refers to both M1/'s, here separated for discussion by A and B designations). However, the total sample from both Harbicht Hill and Mantua Lentil consists of only four specimens (Table 12). From their size alone, these small samples could be interpreted to represent separate species (Figure 17). When larger samples, such as that from Bug Creek Anthills, are included in the analysis, specimens from Mantua Lentil and Harbicht Hill fall within the range of size variation of the Bug Creek Anthills sample. The largest specimens from Bug Creek Anthills are of similar size or larger than those from Mantua Lentil (Figures 17, 18). Therefore, *Stygimys kuszmauli* and "*S. gratus*" are not distinguishable by size. In fact, because of the small samples from Harbicht Hill and Mantua Lentil, the strongest argument for recognizing two species of *Stygimys* on the basis of size would be those from Harbicht Hill and Mantua Lentil, not Bug Creek Anthills and Mantua Lentil.

The cusp formula of M1/'s of *Stygimys* is highly variable. The two M1/'s from Mantua Lentil have cusp formulas of 7:7:5 (PU 14420A, Figure 18) and 7:7:3 (PU 14420B, not figured), each with 1 or 2 cuspules on the anterior part of the internal cusp row. The two M1/'s from Harbicht Hill have cusp formulas of 8-7:8-7:5-6. The large

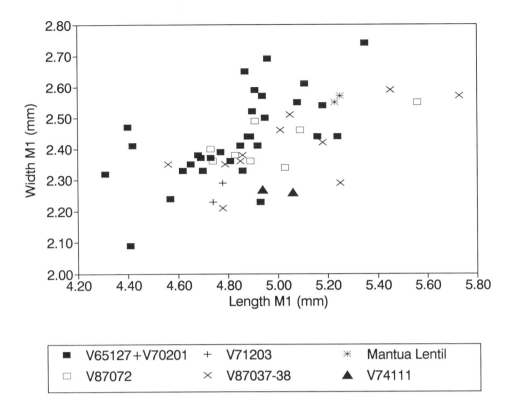

FIGURE 17. Bivariate plot (length vs. width) of M1/'s of *Stygimys* from various localities (see Figure 13).

sample of M1/'s from Bug Creek Anthills exhibits cusp formulas of 8-7:8-7:3-8, with the number on the internal row being highly variable. Those from Bug Creek Anthills usually have 3-5 cusps on the internal row, but many cusp counts depend in part on whether tiny cuspules are counted as cusps. The cusp morphology and formula of M1/'s of *S. kuszmauli* compare closely to those from Mantua Lentil. For example, UCMP 73131 has a cusp formula of 7:7:2 with 4 cuspules (Figure 18), and UCMP 103864 8:8:7 with 1 cuspule (not figured). UCMP 37131 and 103864 are two of the largest M1/'s of *Stygimys* from BCA and are similar in size to those from Mantua Lentil (PU 14420A and 14420B L: 5.23-5.25, W: 2.55-2.57; UCMP 103864 L: 5.35, W: 2.74; UCMP 73131 L: 5.16, W: 2.44). The samples from Bug Creek Anthills and Harbicht Hill are also similar in morphology. Therefore, on the basis of both size and morphology, samples of M1/'s of *Stygimys* from Bug Creek Anthills, Harbicht Hill, and Mantua Lentil are not distinguishable.

M2/: The Mantua Lentil sample does not contain M2/'s. Harbicht Hill M2/'s have a cusp formula of 1-2:3:3-4, and those from Bug Creek Anthills 1-3:3:3-4. A third external cusp is present only on a single M2/ from Bug Creek Anthills. Size measurements and a bivariate length vs. width plot show that the samples are very similar (Table 12, Figure 19).

From this analysis, the samples of *Stygimys* from Bug Creek Anthills, Mantua Lentil, and Harbicht Hill cannot be separated with confidence on the basis of size or

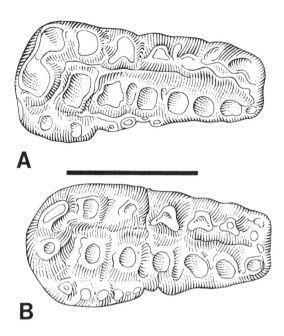

FIGURE 18. *Stygimys kuszmauli* Sloan and Van Valen. (A) Isolated right M1/, UCMP 73131, Bug Creek Anthills (locality V65127), occlusal view. (B) Isolated right M1/, PU 14420A, Mantua Lentil, occlusal view. Scale bar = 4 mm.

morphology. Smaller samples of *Stygimys* (i.e., Mantua Lentil and Harbicht Hill) may exhibit morphological differences, but when larger samples are analyzed (i.e. Bug Creek Anthills), differences appear to have little significance. The morphology of *Stygimys* is highly variable, which can only be fully appreciated with large samples. The Bug Creek Anthills, Mantua Lentil, and Harbicht Hill samples apparently represent a single, highly variable species of *Stygimys* that rightfully retains the name *S. kuszmauli*, the genotypic species.

The sample of *Stygimys* from the upper Hell Creek Formation at McGuire Creek was compared to those in the UCMP collections from Harbicht Hill (V71203) and Bug Creek Anthills (V65127 and V70201), upper Hell Creek Formation, and Worm Coulee 1 (V74111), lower Tullock Formation. Archibald (1982) studied the Worm Coulee 1 sample and referred it to *Stygimys* aff. *S. kuszmauli*. Two of the largest samples from McGuire Creek (V87072, V87037-38) are shown in comparison with Bug Creek Anthills, Harbicht Hill, and Worm Coulee 1 (remeasured for this study) (Figures 13-17, 19). The *Stygimys* samples from all McGuire Creek sites, Harbicht Hill, Bug Creek Anthills, and Worm Coulee 1 exhibit a wide range of morphological variation, but these differences fade when large samples are analyzed. Fossils from these samples are nearly identical in size and morphology (Tables 12, 13, Figures 13-17, 19), and probably belong to a single species. Using these comparisons and the absence of distinguishing characters, I refer the samples from McGuire Creek sites and Worm Coulee 1 to *Stygimys kuszmauli*.

Fox (1989) described a new species of *Stygimys*, *S. cupressus* from the Long Fall

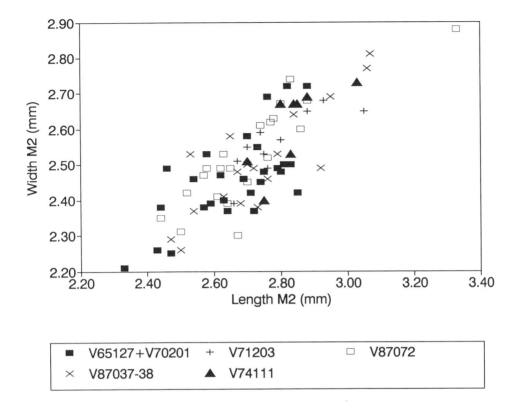

FIGURE 19. Bivariate plot (length vs. width) of M2/'s of *Stygimys* from various localities (see Figure 13).

locality of the Ravenscrag Formation in Saskatchewan. The species is diagnosed by the /P4, whose cutting edge has a symmetrical arc with a long, low, inclined anterior edge leading to the first serration (Fox, 1989). Also, specimens in the University of Alberta collections reveal that the /P4 of *S. cupressus* (n=4) is longer than that of *S. kuszmauli* (n=9) (Fox, 1989). However, lengths of /P4's from Long Fall (4.9mm-5.4mm) fall within the upper part of the range of variation of the large UCMP sample from Bug Creek Anthills (n=34; see Figure 13 and Table 12). A direct comparison of a large sample of *S. kuszmauli* from Bug Creek Anthills and the Long Fall sample of *S. cupressus* would provide an interesting test as to the validity of the latter species because of the highly variable morphology of *S. kuszmauli*.

The morphology of *Stygimys kuszmauli* has been described in detail (Archibald, 1982, formerly *Stygimys* aff. *S. kuszmauli*), but some new information is provided by the fossils from McGuire Creek and Harbicht Hill.

Almost all /P4's from McGuire Creek (n=40) and Harbicht Hill (n=8) have 11 serrations. Three of these from the large McGuire Creek sample have only 10 serrations (UCMP 134739, loc. V87072; UCMP 132666, loc. V87037; UCMP 132803, loc. V87078). The P4/'s from Harbicht Hill and McGuire Creek have a cusp formula of 2-3:6-10:1.

Table 13. Measurements of isolated molars and premolars of *Stygimys kuszmauli* from selected McGuire Creek localities

|  | V87072 | V87098 | V87151 | V87037-8 |
|---|---|---|---|---|
| /P4 length |  |  |  |  |
| Number | 9 | 1 | 4 | 2 |
| Observed Range | 4.69-5.27 | 4.97 | 4.66-5.16 | 4.94-5.34 |
| Mean | 4.88 | 4.97 | 4.85 | 5.14 |
| /P4 Width |  |  |  |  |
| Number | 8 | 2 | 4 | 3 |
| Observed Range | 1.59-2.06 | 1.86-1.90 | 1.79-2.20 | 1.83-2.20 |
| Mean | 1.81 | 1.88 | 1.97 | 1.99 |
| /M1 Length |  |  |  |  |
| Number | 14 | 12 | 10 | 15 |
| Observed Range | 3.92-4.89 | 3.89-4.49 | 3.88-4.56 | 3.98-4.67 |
| Mean | 4.33 | 4.21 | 4.27 | 4.25 |
| /M1 Width |  |  |  |  |
| Number | 14 | 12 | 10 | 15 |
| Observed Range | 1.75-2.03 | 1.73-1.99 | 1.63-2.00 | 1.67-2.09 |
| Mean | 1.88 | 1.88 | 1.83 | 1.88 |
| /M2 Length |  |  |  |  |
| Number | 14 | 8 | 8 | 5 |
| Observed Range | 2.42-2.98 | 2.53-2.85 | 2.47-2.89 | 2.53-2.94 |
| Mean | 2.72 | 2.65 | 2.71 | 2.71 |
| /M2 Width |  |  |  |  |
| Number | 14 | 8 | 8 | 5 |
| Observed Range | 1.86-2.28 | 1.97-2.15 | 1.80-2.13 | 1.87-2.25 |
| Mean | 2.06 | 2.05 | 2.02 | 2.00 |
| P4/ Length |  |  |  |  |
| Number | 8 | 2 | 1 | 11 |
| Observed Range | 3.05-3.47 | 3.34-3.58 | 3.52 | 3.18-3.87 |
| Mean | 3.30 | 3.46 | 3.52 | 3.55 |
| P4/ Width |  |  |  |  |
| Number | 8 | 2 | 1 | 11 |
| Observed Range | 1.47-1.93 | 1.71-1.72 | 1.83 | 1.63-2.21 |
| Mean | 1.78 | 1.72 | 1.83 | 1.84 |
| M1/ Length |  |  |  |  |
| Number | 9 | 6 | 5 | 11 |
| Observed Range | 4.53-5.56 | 4.67-5.26 | 4.62-5.33 | 4.56-5.73 |
| Mean | 4.92 | 4.98 | 4.94 | 5.05 |
| M1/ Width |  |  |  |  |
| Number | 15 | 6 | 8 | 17 |
| Observed Range | 2.24-2.55 | 2.38-2.47 | 2.29-2.54 | 2.21-2.59 |
| Mean | 2.38 | 2.42 | 2.41 | 2.38 |
| M2/ Length |  |  |  |  |
| Number | 21 | 6 | 13 | 19 |
| Observed Range | 2.24-3.33 | 2.54-2.98 | 2.47-2.97 | 2.50-3.07 |
| Mean | 2.71 | 2.77 | 2.74 | 2.74 |
| M2/ Width |  |  |  |  |
| Number | 21 | 6 | 13 | 20 |
| Observed Range | 2.30-2.88 | 2.27-2.69 | 2.30-2.66 | 2.26-2.81 |
| Mean | 2.53 | 2.47 | 2.47 | 2.50 |

McGuire Creek and Harbicht Hill /M2's have a cusp formula of 3-5:2. Ninety percent of the /M2's (n=60) have 4 external cusps. One specimen (UCMP 134725, loc. V87072) has a large fifth external cusp, which is weakly separated from the fourth external cusp by a shallow lingual groove. The M2/'s have a cusp formula of 1-3:3:3-5. Specimens with 4 or 5 internal cusps are rare (5 of 60). When an additional cusp is developed, it is either anterior to and not well separated from the first internal cusp or it is developed posterior to the third internal cusp and deflected medially to occupy a position at the posterior end of the medial valley. One specimen (UCMP 132138, loc. V87151) has cusps developed in both of these positions, giving the tooth 5 cusps in the internal row.

The /M1's have a cusp formula of 6-8:5. When an eighth external cusp is present, it is developed as a robust cusp on a labial bulge within the notch separating cusps 4 and 5, or as a small cusp on a low ridge that extends posteriorly from cusp 4.

The M1/'s of *Stygimys kuszmauli* from McGuire Creek and Harbicht Hill have a cusp formula of 7-8:7-8:3-7. When cusp 8 is present on the external or medial row, it is small and positioned on the anterior margin of the tooth. Three to 7 cusps comprise the internal row. When more than 5 cusps are present, the internal row extends anteriorly to, or slightly past, the midline of the tooth. Cusps on the internal row become progressively smaller anteriorly, with cusps 6 and 7 more appropriately described as tiny cuspules. This is similar to the condition found in M1/'s of *S. kuszmauli* from Mantua Lentil (formerly *S. gratus*).

Family TAENIOLABIDIDAE Granger and Simpson, 1929
*Catopsalis* Cope, 1882a

*Comments*: *Catopsalis* is a paraphyletic taxon (Simmons and Desui, 1986) known from Upper Cretaceous strata in Asia (Kielan-Jaworowska and Sloan, 1979) and Puercan to Tiffanian strata in North America (Archibald et al., 1987; Archibald and Lofgren, 1990). Two species of the genus, *C. joyneri* and *C. alexanderi*, have been reported from eastern Montana (Sloan and Van Valen, 1965; Middleton, 1982). *C. joyneri* is best known from the Bug Creek sequence of channel fills (Bug Creek Anthills, Bug Creek West, Harbicht Hill) in the upper Hell Creek Formation of McCone County, Montana. The type of *C. alexanderi* comes from the Alexander locality in the Denver Formation, Arapahoe County, Colorado. The hypodigm of *C. alexanderi* included specimens originally referred to *C. foliatus* (see Middleton, 1982, p. 1198). Two specimens also referred to *C. alexanderi* in the hypodigm were collected from the Tullock Formation of Garfield County, Montana. These fossils (UCMP 116954, M1/ fragment, UCMP loc. V74111; UCMP 124404, M2/, UCMP loc. V74110) were originally referred to *Catopsalis* cf. *C. foliatus* (see Archibald, 1982).

Species of *Catopsalis* display certain evolutionary trends when set in decreasing chronologic order (Kielan-Jaworowska and Sloan, 1979; Middleton, 1982). One of these is a gradual increase in size: *Catopsalis joyneri* is smaller than *C. alexanderi*, which in turn is smaller than *C. foliatus*.

*Catopsalis joyneri*, and the distinctly larger *C. alexanderi*, are both present at McGuire Creek, but never at the same locality. This might be a factor of the small sample size (n=10), or may reflect temporal differences between sites.

*Catopsalis joyneri* Sloan and Van Valen, 1965
Table 14

*Catopsalis joyneri* Sloan and Van Valen, 1965, p. 225.

*Type.* UMVP 1494, right maxilla with complete palate, M1/, roots of P4/, alveoli of P3/ and M2/.

*Type locality.* Bug Creek Anthills, Hell Creek Formation.

*Referred specimens.* One P4/ from loc. V87071. One M2/, 1 /M1 from loc. V87151. One M2/ from loc. V87153.

*Localities.* UCMP locs. V87071, V87151, and V87153.

*Distribution.* Upper Hell Creek Formation, Montana (Puercan); possibly upper Frenchman and Ravenscrag formations, Saskatchewan (both Puercan).

*Description.* In 1965, Sloan and Van Valen published a brief description of *C. joyneri* based on fossils collected at Bug Creek Anthills, McCone County, Montana. A more complete description of the species has yet to appear. Therefore, a brief description of the small McGuire Creek sample (n=4) is given below.

Table 14. Measurements of isolated molars and premolars of *Catopsalis joyneri* from McGuire Creek sites and Bug Creek Anthills (V65127)

| McGuire Creek Locality | Specimen | Tooth Site | Length | Width |
|---|---|---|---|---|
| V87071 | 133167 | P4/ | 3.21 | 2.31 |
| V87153 | 133567 | M2/ | 5.79 | 4.98 |
| V87151 | 132180 | M2/ | — | 4.62 |
| V87151 | 132181 | /M1 | 5.93 | 3.34 |

| Bug Creek Anthills | Number | Observed Range | Mean |
|---|---|---|---|
| P4/ Length | 6 | 2.98-3.47 | 3.22 (3.3)[b] |
| Width | 6 | 2.19-2.35 | 2.28 |
| M1/ Length | 11 | 7.56-8.65 | 8.09 (8.1)[b] |
| Width | 11 | 4.15-4.70 | 4.38 |
| M2/ Length | 4 | 5.0-5.8[a] | 5.5[a] (5.2)[b] |
| Width | 4 | 4.5-4.8[a] | 4.6[a] |
| /M1 Length | 13 | 5.71-7.17 | 6.44 (7.0)[b] |
| Width | 13 | 2.82-3.63 | 3.33 |
| /M2 Length | 4 | 5.74-6.10 | 5.91 (5.5)[b] |
| Width | 4 | 4.00-4.27 | 4.10 |

a. data from Archibald, 1982, p. 71.
b. length from diagnosis by Sloan and Van Valen, 1965, p. 225.

Dental measurements of *C. joyneri* from McGuire Creek are presented in Table 14. The UCMP sample of *C. joyneri* from Bug Creek Anthills was measured, and these data are also given in Table 14 for comparative purposes.

The single P4/ (UCMP 133167) referable to *C. joyneri* is heavily worn on its medial crest, making determination of the cusp formula impossible (1:5?:?). The single external cusp is small and positioned on a well developed cingulum on the posterolabial edge of the tooth. The medial row probably had 5 cusps, but only cusps 1-3 remain. The medial row appears to be similar to that of the larger species of *Catopsalis, C. alexanderi* (see Middleton, 1982, p. 1201). The posterolingual part of the tooth is heavily worn and any trace of internal cusps has been removed. A short but distinct cingulum is present labial to medial cusps 4-5?.

A moderately worn /M1 (UCMP 132181) has a cusp formula of 5:4 and is similar in size and morphology to /M1's in the UCMP sample from Bug Creek Anthills. UCMP 133567, a slightly worn M2/, has a cusp formula of 1:3:3. Its internal cusps are subequal in size. In the medial row, the anterior cusp is small and transversely elongated, and the posterior cusp is the largest. The single external cusp, on the anterolabial edge of the tooth, is narrow and ridge-like, with its long axis oriented anterolingual-posterolabial. This cusp is positioned on the anterolabial edge of the tooth. The other M2/, UCMP 132180, is heavily worn and offers no additional morphological information.

*Catopsalis alexanderi* Middleton, 1982

Table 15

*Catopsalis alexanderi* Middleton, 1982, p. 1198.

  *Type.* UCM 34979, right lower jaw with I, /P4, /M1-2.

  *Type locality.* Alexander locality (UCM 77267), Denver Formation, Colorado.

  *Referred specimens.* One M2/, UCMP 136095, from loc. V71203. One /M2, UCMP 132478, from loc. V87034. One M1/, UCMP 132659, from loc. V87037.

  *Localities.* UCMP locs. V71203, V87034, and V87037.

  *Distribution.* Denver Formation, Colorado; Polecat Bench Formation, Wyoming; upper Hell Creek and lower Tullock formations, Montana (all Puercan).

  *Description.* The three specimens referred to *C. alexanderi* are similar in size to those in Middleton's (1982, tables 1, 2) sample and are significantly larger than those of *C. joyneri*

Table 15. Measurements of isolated molars of *Catopsalis alexanderi* from McGuire Creek localities and Harbicht Hill (V71203)

| Locality | Specimen | Tooth Site | Length | Width |
|---|---|---|---|---|
| V71203 | 136095 | M2/ | 6.87 | 5.66 |
| V87034 | 132478 | /M2 | 6.76[a] | 4.90 |
| V87037 | 132659 | M1/ | 9.15 | 4.91 |

a. Approximate measurement.

from Bug Creek Anthills (Tables 14, 15). Morphologically, these specimens are similar to those described by Middleton (1982), and only a few comments are necessary.

UCMP 132659, an unworn M1/, is missing the anterior portion of the external cusp row because of breakage, and only 5 cusps are preserved. However, a proportional comparison to complete M1/'s of *C. joyneri* reveal that UCMP 132659 probably had at least 7 cusps. The middle and internal rows have 8 cusps each. The internal row extends anteriorly about 90% of the total length of the tooth. UCMP 132478, a moderately worn /M2, is missing the extreme posterolabial corner of the tooth. From comparisons to complete /M2's of *C. joyneri* from Bug Creek Anthills, it appears that the length of UCMP 132478 (Table 15) is close to the original maximum length. This specimen has a cusp formula of 3:2+. The internal row has a small narrow anterio-posteriorly orientated cuspule on the postero-labial margin of the tooth.

The small UCMP collection from Harbicht Hill (V71203) contains one isolated tooth, an M2/ (UCMP 136095), referable to *Catopsalis* (Table 15); it is much larger than that of *C. joyneri* (Table 14). Based on size and morphology (cusp formula 1:3:3), UCMP 136095 is referable to *C. alexanderi*.

The presence of *C. alexanderi* at Harbicht Hill and McGuire Creek are new records of the species in the upper Hell Creek Formation. Previously, only *C. joyneri* was listed as present at Harbicht Hill (Sloan and Van Valen, 1965). Because *C. alexanderi* was described more recently (Middleton, 1982) than Sloan and Van Valen's (1965) Harbicht Hill faunal list, it is possible that either both species or only *C. alexanderi* is present at Harbicht Hill.

*Catopsalis* sp. indet.

Table 16

*Referred specimens.* One /I from loc. V87033. One /I from loc. V87035. One /I fragment from loc. V87070. One /I fragment from loc. V87071. One /Mx fragment from loc. V87074.

*Localities.* UCMP locs. V87033, V87035, V87070, V87071, and V87074.

*Description.* These four lower incisors and a partial molar are referable to *Catopsalis* with reasonable confidence, but are too fragmentary for identification to species level.

Lower incisors of *Catopsalis alexanderi* and *C. joyneri* are distinguished by size (Middleton, 1982). Lower incisors of *Catopsalis* from McGuire Creek appear to be referable to *C. alexanderi* as shown by comparison of measurements from Table 16 and Table 2

Table 16. Measurements of lower incisors of *Catopsalis* sp. indet. from McGuire Creek localities

| Locality | Specimen | Maximum Width | Maximum Depth |
|----------|----------|---------------|---------------|
| V87033 | 132431 | 3.81 | 5.38 |
| V87035 | 132539 | 3.50[a] | 4.90[a] |
| V87071 | 134603 | 3.41 | 5.49 |

a: Approximate measurement.

of Middleton (1982, p. 1201). However, the size range and mean dimensions of incisors referred to *C. joyneri* were not presented in the original diagnosis of the species (Sloan and Van Valen, 1965). Therefore, the incisors listed in Table 16 are not referred to *C. alexanderi*, although this possibility appears most likely, based on size.

Suborder, incertae sedis

Family CIMOLOMYIDAE (Marsh, 1889b)

Sloan and Van Valen, 1965

*Meniscoessus* Cope, 1882b

*Meniscoessus robustus* (Marsh, 1889a) Osborn, 1891

Table 17

*Dipriodon robustus* Marsh, 1889a, p. 85 (see Archibald, 1982, p. 75, for synonymies).

*Type.* YPM 11234, right /M2 (Marsh 1889a, pl. II, figs. 13-15).

*Type locality.* Mammal locality no. 2 of Lull (1915), UCMP loc. V5815, Lance Formation, Wyoming.

---

Table 17. Measurements of isolated premolars and molars of *Meniscoessus robustus* from Lancian (V85092) and Puercan localities at McGuire Creek

| Tooth Site | V85092 Lancian Observed | | | Puercan Localities Observed | | |
|---|---|---|---|---|---|---|
| | N | Range | Mean | N | Range | Mean |
| **/P4** | | | | | | |
| Length | 1 | 7.09 | 7.09 | – | — | — |
| Width | 1 | 3.46 | 3.46 | – | — | — |
| **/M1** | | | | | | |
| Length | 3 | 7.54-8.12 | 7.76 | 4 | 7.69-8.72 | 8.29 |
| Width | 3 | 3.48-4.18 | 3.93 | 5 | 3.66-4.37 | 4.04 |
| **/M2** | | | | | | |
| Length | – | — | — | 4 | 6.40-7.70 | 7.22 |
| Width | 1 | 4.33 | 4.33 | 4 | 3.96-4.77 | 4.47 |
| **P4/** | | | | | | |
| Length | – | — | — | 2 | 4.25-4.42 | 4.34 |
| Width | – | — | — | 3 | 2.96-3.28 | 3.11 |
| **M1/** | | | | | | |
| Length | 2 | 9.05-9.79 | 9.42 | 1 | 9.82 | 9.82 |
| Width | 2 | 5.40-5.42 | 5.41 | 1 | 5.75 | 5.75 |
| **M2/** | | | | | | |
| Length | 2 | 6.85-7.52 | 7.19 | 7 | 6.75-8.39 | 7.40 |
| Width | 2 | 5.91-6.59 | 6.25 | 7 | 5.75-7.28 | 6.44 |

*Referred specimens.* Four M1/'s, 4 M2/'s, 1 /P4, 3 /M1's, 1 /M2 from loc. V85092. Three I's, 3 P4/'s, 1 M1/, 7 M2/'s, 2 /P4 fragments, 7 /M1's, 5 /M2's, and 2 Mx fragments from other UCMP localities.

*Localities.* UCMP locs. V84151, V85092, V86031, V87030, V87034, V87035, V87037, V87038, V87071, V87072, V87074, V87084, V87098, V87151, and V88037.

*Revised diagnosis.* Archibald, 1982, p. 75.

*Distribution.* St. Mary River Formation, Alberta (Edmontonian); Hell Creek Formation, South Dakota, and Lance Formation, Wyoming (both Lancian); Frenchman Formation, Saskatchewan (Lancian, in part); Hell Creek Formation, Montana (Lancian and *Puercan); possibly upper Frenchman Formation and Ravenscrag Formation, Saskatchewan (both *Puercan).

*Description.* The morphology of *Meniscoessus robustus* has been adequately described by Clemens (1964) and Archibald (1982). McGuire Creek specimens referred to *M. robustus* are similar in size and morphology to fossil material described in those works, and no additions are necessary. Dental measurements are given in Table 17.

*Comments.* Fossils referred to *Meniscoessus robustus* come from both Lancian and Puercan localities in the upper Hell Creek Formation at McGuire Creek. All the Lancian specimens are from UCMP locality V85092. The remaining teeth were recovered from 14 widely scattered Puercan localities. The presence of *Meniscoessus* at these Puercan sites probably is the result of reworking. Alternatively, the presence of two morphologically distinct samples, one each from Puercan and Lancian sites, would suggest that *Meniscoessus* might have survived into the Puercan. To test this hypothesis, the entire McGuire Creek sample of *Meniscoessus* was measured and then divided into the two bulk age groupings (Lancian and Puercan). Comparison of these admittedly small samples gave no indication of significant size or morphological differences (Table 17). The Puercan specimens are, on average, slightly larger, but almost all of these fossils fall within the range of variation displayed by *M. robustus* from the Lance Formation of Wyoming (Clemens, 1964, table 10) and the Lancian part of the Hell Creek Formation of Montana (Archibald, 1982, table 13). The size of dental elements of *M. robustus* appears to be highly variable, and until collection of a larger sample from McGuire Creek shows evidence of a distinct bimodality between Lancian and Puercan specimens of *Meniscoessus*, it appears only that one species is present.

Family ?CIMOLOMYIDAE
*Essonodon* Simpson, 1927a
Table 18; Figure 20

*Essonodon browni* Simpson, 1927a, p. 2.

*Type.* AMNH 14410, an isolated right /M2 (Simpson, 1927a, fig. 1).

*Type locality.* Near the head of the East Fork of Crooked Creek, Hell Creek Formation, Montana.

*Referred specimen.* One /M2, UCMP 132182, from loc. V87151.

*Locality.* UCMP loc. V87151.

*Distribution.* Kirkland Formation, New Mexico (Lancian?); Hell Creek Formation,

Table 18. Measurements of /M2's of *Essonodon browni* from various localities[a]

| Locality | Specimen | Length | Width | Length/Width |
|---|---|---|---|---|
| V87151 | UCMP 132182 | 2.49 | 3.00 | .83 |
| V73083 | UCMP 108965 | 2.58 | 2.71 | .95 |
| Crooked Creek, Montana (holotype) | AMNH 14410 | 2.45 | 3.61 | .68 |
| Lance Formation, Wyoming | YPM 14907 | 2.39 | 3.23 | .74 |

a. All data from Archibald (1982) except UCMP 132182. UCMP 108965 and AMNH 14410 are from the Hell Creek Formation, Garfield County, Montana. UCMP 132182 is from McGuire Creek, McCone County, Montana.

Montana (Lancian and *Puercan); possibly Frenchman Formation, Saskatchewan (Lancian, in part).

*Revised diagnosis.* Archibald, 1982, p. 89-90.

*Description.* The single tooth referred to *Essonodon browni* from McGuire Creek is a moderately worn /M2 (UCMP 132182) with a cusp formula of 3:2. This specimen is similar in size and morphology to the holotype (AMNH 14410, an isolated /M2) and to the few other /M2's known of the species (Table 18). UCMP 132182 (Figure 20) differs slightly from the holotype in development and orientation of the posterior external cusp, which is smaller, located more posteriorly, and situated on the extreme posterolabial edge of the tooth. A well-developed, transversely oriented ridge, which is enhanced by the moderate state of wear of the tooth, forms the posterior margin of the tooth and connects the external and internal cusps. On the holotype, this ridge is less marked and the posterior external cusp is larger and situated more anteriorly (Simpson, 1927a).

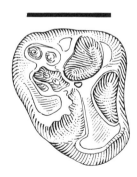

FIGURE 20. *Essonodon browni* Simpson. Isolated left /M2, UCMP 132182, locality V87151, occlusal view. Scale bar = 2 mm.

Suborder and Family, incertae sedis
*Cimexomys* Sloan and Van Valen, 1965
*Cimexomys minor* Sloan and Van Valen, 1965
Tables 19-20

*Cimexomys minor* Sloan and Van Valen, 1965, p. 221.

*Type.* SPSM 62-2115, left mandible with /P3-4, and alveoli for /I, /M1, and /M2 (Sloan and Van Valen, 1965, fig. 2).

*Type locality.* Bug Creek Anthills, Hell Creek Formation, Montana.

*Referred specimens.* One /P4 from loc. V87030. Three M1/'s, 3 /P4's, from loc. V87037. Eight M1/'s, 1 /P4, 2 P4/'s from loc. V87038. One M1/ from loc. V87051. One M1/ fragment from loc. V87071. Seven M1/'s, 5 /P4's, 2 P4/'s from loc. V87072. Four M1/'s, 2 /P4's from loc. V87074. One M1/ from loc. V87077. One P4/ from loc. V87091. One M1/, 2 /P4's from loc. V87098.

*Localities.* UCMP locs. V87030, V87037, V87038, V87051, V87071, V87072, V87074, V87077, V87091, and V87098.

*Distribution.* Lance Formation, Wyoming (Lancian); Hell Creek Formation, Montana (Lancian and Puercan); lower Tullock Formation, Montana, and Ravenscrag Formation, Saskatchewan (both Puercan).

*Description.* The McGuire Creek sample of *Cimexomys minor* is limited to /P4's, P4/'s, and M1/'s. The abundant presence of *C. minor* at some sites (especially locs. V87072, V87037, and V87038) would indicate that /M1's, /M2's, and M2/'s are undoubtedly also present in the collections from these and other localities at McGuire Creek. Unfortunately, current knowledge does not allow teeth of *C. minor* to be distinguished from smaller species of *Mesodma*, unless found in association with /P4's, P4/'s, or M1/'s (Archibald, 1982). Therefore, the large *Mesodma* sample from McGuire Creek (n>2200) must include a total number of /M1's, /M2's, and M2/'s of *Cimexomys minor* roughly equal to the number of identifiable /P4's, P4/'s, and M1/'s (n=45).

Specimens referred to *Cimexomys minor* are similar to those described by Sloan and Van Valen (1965), Clemens (1964, 1973), and Archibald (1982), and only a few comments are necessary. M1/'s from McGuire Creek have a cusp formula of 4-5:6-7:1. Twenty percent (n=5) of them have 7 cusps in the medial row, and four of these teeth also have 5 cusps in the external row. The internal cusp is elongated antero-posteriorly and forms a short ridge, which does not extend to the midline of the tooth. In a few unworn specimens, this internal ridge can be subdivided into 2 or 3 weakly developed cuspules. /P4's have 8 serrations with 5-6 internal and external ridges, which are not developed on the first or last serrations. All P4/'s from McGuire Creek have a cusp formula of 3:5:2.

Dental measurements of *C. minor* from the localities which form the majority of the McGuire Creek sample are presented in Table 19. Fossils from McGuire Creek are similar in size to those from the Lance Formation of Wyoming (UCMP loc. V5711) (Clemens, 1973, table 8) and the lower Tullock Formation of Montana (UCMP loc. V74111) (Archibald, 1982, table 15). M1/'s in the UCMP collections from the type locality of the species (Bug Creek Anthills, UCMP loc. V65127) also compare closely in size to those from McGuire Creek (Table 20).

Table 19.  Measurements of isolated M1/'s, P4/'s, and /P4's of *Cimexomys minor* from four McGuire Creek localities

| UCMP Locality: | V87072 | V87074 | V87037-38[a] |
|---|---|---|---|
| **M1/ Length** | | | |
| Number | 7 | 4 | 10 |
| Observed Range | 2.31-2.66 | 2.36-2.57 | 2.32-2.68 |
| Mean | 2.51 | 2.48 | 2.51 |
| **M1/ Width** | | | |
| Number | 7 | 4 | 11 |
| Observed Range | 1.30-1.41 | 1.29-1.46 | 1.22-1.43 |
| Mean | 1.36 | 1.39 | 1.36 |
| **/P4 Length** | | | |
| Number | 5 | 2 | 3 |
| Observed Range | 2.70-2.77 | 2.52-2.69 | 2.74-2.86 |
| Mean | 2.73 | 2.61 | 2.81 |
| **/P4 Width** | | | |
| Number | 5 | 2 | 3 |
| Observed Range | .85-1.04 | .90-.90 | .92-.96 |
| Mean | .92 | .90 | .94 |
| **P4/ Length** | | | |
| Number | 2 | – | 2 |
| Observed Range | 2.09-2.25 | — | 2.24-2.27 |
| Mean | 2.17 | — | 2.26 |
| **P4/ Width** | | | |
| Number | 2 | – | 2 |
| Observed Range | 1.13-1.25 | — | 1.19-1.47 |
| Mean | 1.19 | — | 1.33 |

a: V87037 and V87038 are virtually the same site.

Table 20.  Measurements of isolated M1/'s of *Cimexomys minor* from Bug Creek Anthills (V65127)

| M1/ | Number | Observed Range | Mean |
|---|---|---|---|
| Length | 11 | 2.41-2.69 | 2.57 |
| Width | 10 | 1.38-1.47 | 1.43 |

*Cimexomys gratus* Jepsen, 1930
Tables 21-22; Figures 21-22

*Cimexomys hausoi* Archibald, 1982, p. 105.

*Type.* PU 13373, incomplete left ramus with /M2, and alveoli of I, /P4, and /M1 (Jepsen, 1930, pl. IV, fig. 8).

*Type locality.* Mantua Lentil, Polecat Bench Formation (Fort Union Formation), Wyoming.

*Referred specimens.* One /P4, 4 P4/'s, 3 /M1's, 2 M1/'s, 2 /M2's, 1 M2/, from loc. V87037. One /P4, 1 /M1, 2 M1/'s, 5 M2/'s, from loc. V87035. Three P4/'s, 2 /P4's, 3 /M1's, 6 M1/'s, 3 /M2's, from loc. V87038. Two P4/'s, 4 /M1's, 1 M1/, 2 /M2's, 6 M2/'s, from loc. V87151. One P4/, 1 /M1, 1 M1/, 3 M2/'s, from loc. V87051. One P4/, 1 /P4, 1 /M1, 6 M1/'s, 2 M2/'s, from loc. V87098. One /P4, 2 /M1's, 2 M1/'s, 3 /M2's, 4 M2/'s, from loc. V87071. One /P4, 4 P4/'s, 1 /M1, 7 M1/'s, 7 M2/'s, 3 /M2's, from loc. V87074. Four /P4's, 8 P4/'s, 7 /M1's, 10 M1/'s, 5 /M2's, 8 M2/'s, from loc. V87072. One /M1, 1 M1/, from loc. V65127. Two M1/'s from loc. V71203. Seven isolated premolars and molars from other UCMP localities.

*Localities.* UCMP locs. V65127, V71203, V87028, V87030, V87035, V87037, V87038, V87051, V87052, V87071, V87072, V87074, V87077, V87098, V87101, V87114, and V87151.

*Distribution.* Upper Hell Creek and lower Tullock formations, Montana (both Puercan); Polecat Bench Formation (Fort Union Formation), Wyoming, and possibly Ravenscrag Formation, Saskatchewan (both Puercan).

*Discussion.* During the course of study of the *Stygimys gratus* sample from Mantua Lentil, which was originally described by Jepsen (1930, 1940), it was apparent that the holotype and the remainder of the hypodigm represented two different taxa. The hypodigm, excluding the holotype, is indistinguishable from *Stygimys kuszmauli* from Bug Creek Anthills and is referred to that species (see discussion of *S. kuszmauli*). The holotype of *S. gratus*, a fragmentary dentary with a heavily worn /M2 (Figure 21), is similar to the holotype of *Cimexomys hausoi* from Worm Coulee 1 (V74111), lower Tullock Formation, Montana (Archibald, 1982, figure 36), where comparisons can be made. Measurement of the depth of the dentary below the /M2 (measured from alveolar rim of the /M2 to ventral edge of the dentary) indicates that the holotype of *Stygimys gratus* (PU 13373: 4.2mm) is similar to the holotype of *Cimexomys hausoi* (UCMP 117000: 3.9mm). These two specimens are much smaller than the two other dentary fragments from Mantua Lentil referred to *Stygimys gratus* (PU 14418: 7.7mm; PU 14419: 6.5mm). Also, the /M2's of these holotypes are similar in size (*Cimexomys hausoi* UCMP 117000: L: 2.32, W: 1.86; Archibald, 1982, table 16; *Stygimys gratus* PU 13373: L: 2.37, W: 1.85) and have cusp formulas of 4:2. However, based on size, PU 13373 falls within the range of variation of /M2's of *S. kuszmauli* from Bug Creek Anthills (L: 2.38-2.96; W: 1.79-2.22). Therefore, comparison of PU 13373 to worn /M2's of *Cimexomys hausoi* and *Stygimys kuszmauli* from their respective type localities is needed to eliminate the possibility that PU 13373 represents a small individual of the latter species.

The /M2 of PU 13373 is similar to worn /M2's of *Cimexomys hausoi* from Worm Coulee 1 (V74111) in the following ways: (1) /M2's of *Stygimys kuszmauli* initially exhibit the heaviest wear on the internal cusp row, while those of *Cimexomys hausoi*

Table 21. Measurements of isolated M1/'s of *Cimexomys gratus* from Bug Creek Anthills (locs. V65127 and V70201, a sublocality of V65127), Harbicht Hill (V71203), Worm Coulee 1 (V74111), and Up-Up-the-Creek 2 and 3 (V87037-38)[a]

| Locality | UCMP Specimen | Length | Width |
|----------|---------------|--------|-------|
| V65127 | 73870 | 3.05 | 1.82 |
| V65127 | 104897 | 3.18 | 2.02 |
| V65127 | 105451 | 3.08 | 1.88 |
| V70201 | 92553 | 2.99 | 1.72 |
| V70201 | 92551 | 3.08 | 1.78 |
| V70201 | 98130 | 3.04 | 1.70 |
| V71203 | 136096 | 3.29 | 1.90 |
| V71203 | 136097 | 3.39 | 1.91 |
| V74111 | 117013 | 3.08 | 1.79 |
| V74111 | 117015 | 3.54 | 2.14 |
| V74111 | 117016 | 3.30 | 1.97 |
| V74111 | 117017 | 3.49 | 1.96 |
| V74111 | 137265 | 3.22 | 1.98 |
| V87037 | 132698 | 3.10 | 1.89 |
| V87037 | 132700 | 2.98 | 1.73 |
| V87038 | 133040 | 3.19 | 2.06 |
| V87038 | 133042 | 3.53 | 2.05 |
| V87038 | 133043 | 3.20 | 1.84 |

a. All localities from the Hell Creek Formation of McCone County, Montana, except V74111, which is from the Tullock Formation, Garfield County, Montana.

exhibit heaviest wear on the external cusp row first (Archibald, 1982). (2) Transverse grooves between cusps of *Stygimys kuszmauli* are wide and broad, especially those between the internal cusps and the two anterior external cusps. /M2's of *Cimexomys hausoi* have narrow deep grooves between cusps on both cusp rows, which are deepest medially on each row. (3) The internal cusps of *Stygimys kuszmauli* merge to form a longitudinal ridge only after extreme wear, because a deep, broad groove separates the two cusps. After moderate wear, the transverse grooves between cusps of *Cimexomys hausoi* are removed from the lingual edge of the internal row and labial edge of the external row, and cusps in each row merge to form two longitudinal ridges. (4) An irregular series of deep pits is present in the enamel on the occlusal surface of /M2's of *Cimexomys hausoi*. Pitting of the enamel is poorly developed or absent in *Stygimys kuszmauli*. (5) The anterior end of /M2's of *Cimexomys hausoi* is flattened and has a straight anterior face to which the posterior end of the /M1 abuts. In contrast, the first internal cusp of *Stygimys kuszmauli* is swollen and rounded, and the anterior face is not flattened to form a straight anterior surface.

The holotypes of *Stygimys gratus* and *Cimexomys hausoi* are virtually identical, and these specimens represent a single species. Neither *Stygimys gratus* nor *Cimexomys hausoi* are the genotypic species of their respective genera, and both genera were described

Table 22. Measurements of isolated premolars and molars of *Cimexomys gratus* from McGuire Creek localities

| UCMP Locality: | V87072 | V87037-38 | Entire McGuire Creek Sample |
|---|---|---|---|
| P4/ Length | | | |
| Number | 8 | 5 | 15 |
| Observed Range | 2.77-3.33 | 2.98-3.25 | 2.77-3.38 |
| Mean | 3.08 | 3.08 | 3.13 |
| P4/ Width | | | |
| Number | 8 | 5 | 15 |
| Observed Range | 1.34-1.81 | 1.40-1.91 | 1.31-1.91 |
| Mean | 1.68 | 1.70 | 1.66 |
| /P4 Length | | | |
| Number | 3 | 2 | 10 |
| Observed Range | 3.78-4.23 | 3.74-4.24 | 3.56-4.24 |
| Mean | 4.01 | 3.99 | 3.88 |
| /P4 Width | | | |
| Number | 3 | 3 | 11 |
| Observed Range | 1.33-1.41 | 1.35-1.61 | 1.18-1.61 |
| Mean | 1.37 | 1.44 | 1.35 |
| M1/ Length | | | |
| Number | 6 | 5 | 26 |
| Observed Range | 3.13-3.60 | 2.98-3.53 | 2.98-3.60 |
| Mean | 3.32 | 3.20 | 3.29 |
| M1/ Width | | | |
| Number | 10 | 5 | 33 |
| Observed Range | 1.72-2.03 | 1.73-2.06 | 1.72-2.08 |
| Mean | 1.90 | 1.91 | 1.87 |
| /M1 Length | | | |
| Number | 7 | 5 | 20 |
| Observed Range | 3.06-3.99 | 2.94-3.60 | 2.94-3.99 |
| Mean | 3.46 | 3.22 | 3.35 |
| /M1 Width | | | |
| Number | 7 | 5 | 22 |
| Observed Range | 1.37-1.97 | 1.39-1.67 | 1.37-1.97 |
| Mean | 1.66 | 1.52 | 1.58 |
| M2/ Length | | | |
| Number | 7 | 2 | 31 |
| Observed Range | 2.02-2.46 | 2.30-2.34 | 1.98-2.50 |
| Mean | 2.21 | 2.32 | 2.21 |
| M2/ Width | | | |
| Number | 7 | 3 | 32 |
| Observed Range | 1.88-2.25 | 2.02-2.15 | 1.74-2.26 |
| Mean | 2.03 | 2.10 | 2.05 |

| UCMP Locality: | V87072 | V87037-38 | Entire McGuire Creek Sample |
|---|---|---|---|
| /M2 Length | | | |
| Number | 4 | 3 | 16 |
| Observed Range | 2.02-2.25 | 1.92-2.31 | 1.92-2.40 |
| Mean | 2.10 | 2.12 | 1.12 |
| /M2 Width | | | |
| Number | 5 | 3 | 18 |
| Observed Range | 1.61-1.84 | 1.66-1.91 | 1.58-1.91 |
| Mean | 1.70 | 1.81 | 1.74 |

in the same work (Sloan and Van Valen, 1965). Thus, the holotype of *Stygimys gratus* (PU 13373) replaces the holotype of *Cimexomys hausoi* (UCMP 117000) and becomes the namebearer of that species, now recognized as *Cimexomys gratus*.

During the measurement of M1/'s of *Cimexomys* from Bug Creek Anthills, it became evident that there were two species present in the sample. Most specimens are small and can be referred to *C. minor* (see Table 20), but six teeth are much larger. Based on cusp and overall tooth morphology, these six teeth are indistinguishable from those referred to *C. gratus* from Worm Coulee 1 (V74111). Based on size, most fall within the range of variation displayed by *C. gratus* (Table 21, Figure 22). The specimens from Bug Creek Anthills form a cluster near the two smallest specimens from Worm Coulee 1, the type locality of *C. gratus* (Figure 22). Specimens from Bug Creek Anthills would presumably be referred to *C. gratus* if they were collected from Worm Coulee 1. As an example, the range in variation of the sample of *C. gratus* from V87037-38 is nearly as large as both the Bug Creek Anthills and Worm Coulee 1 samples in combination (Figure 22, Table 21). In any case, evidently *C. gratus* and *C. minor* are both present at Bug Creek Anthills. In Sloan and Van Valen's (1965) brief description of the fauna at Bug Creek Anthills, no mention is made of a larger species of *Cimexomys*.

Although improbable, a third species of *Cimexomys* might be present in the upper Hell Creek Formation of McCone County, Montana. In the original description of *C. minor*, Sloan and Van Valen (1965) reported a second and larger species of *Cimexomys* from Harbicht Hill and referred a single specimen (PU 14999) from Mantua Lentil (Polecat Bench Formation, Wyoming) to it. However, they did not name or describe the species. Subsequently, Archibald (1982) described a large species, *C. gratus* (formerly *C. hausoi*), from the lower Tullock Formation of Garfield County, Montana.

Comparison of *C. gratus* to the unnamed species at Harbicht Hill was not possible, because the original specimens from Harbicht Hill could not be located (Archibald, 1982). Examination of the specimen from Mantua Lentil (PU 14999) by Middleton (1983) and comparison with Archibald's description and measurements of *C. gratus* revealed no significant differences (Middleton, 1983, p. 150-152). Also, the small UCMP collection from Harbicht Hill (V71203) contains two M1/'s (UCMP 136096 and 136097) referable to *Cimexomys*. These teeth were compared to M1/'s of *C. gratus* from Worm Coulee 1 (V74111) and found to be morphologically indistinguishable (see Table 21 and

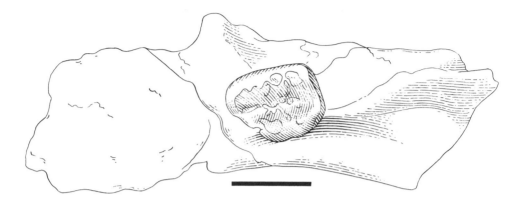

FIGURE 21. *Cimexomys gratus* Jepsen. Left dentary fragment with /M2, and alveoli of /I, /P4, and /M1, PU 13373 (holotype), Mantua Lentil, occlusal view. Scale bar = 2 mm.

Figure 22; both samples have cusp formulas of 5:6:1). Therefore, the unnamed large species of *Cimexomys* at Harbicht Hill mentioned by Sloan and Van Valen (1965) almost certainly is *C. gratus.*

*Description.* Fossils of *C. gratus* from McGuire Creek are similar to those described by Archibald (1982). P4/'s from McGuire Creek have a cusp formula of 3-4:5-6:2-3, with 3 internal cusps present in only one specimen. /P4's have 8-9 serrations, with the majority (9 of 11) having nine. Specimens with 8 serrations have 6 internal and external ridges, none of which are developed from the first or last serration. Fossils with 9 serrations exhibit more variation in ridge development with 6-8 external and internal ridges present. The ultimate serration never has a ridge. In specimens with 6 external or internal ridges, the penultimate serration lacks a ridge. Three specimens have a ridge count of 8 (external and internal), because the first serration has a very short, weakly developed ridge.

M1/'s have a cusp formula of 4-5:6:1(R-4). An anteriorly positioned fifth external cusp is well developed when present (65% of 40 teeth), and all specimens have 6 medial cusps. /M1's have a cusp formula of 6-7:4, with 6 external cusps present in only a single specimen. M2/'s have a cusp formula of 1:3:4-5, with a single tooth having a fifth internal cusp. /M2's have a cusp formula of 4:2. Dental measurements are given in Table 22.

Archibald (1982), analyzing fossils from the Worm Coulee 1 (UCMP loc. V74111) locality in the lower Tullock Formation of Garfield County, Montana, noted that most of the differences between *C. minor* and *C. gratus* (formerly *C. hausoi*) are attributable to size. This is also true for the McGuire Creek sample of the genus, although three morphological differences are worth noting. A few M1/'s of *C. minor* from McGuire Creek have 7 medial cusps instead of six. M1/'s of *C. gratus* from McGuire Creek always have 6 medial cusps. Also, in many M1/'s of *C. gratus* the fifth anterior external cusp is large and well separated from the fourth. When a fifth anterior external cusp is present in *C. minor*, it is small and not well separated from the fourth external cusp. /P4's of *C. minor* have 8 serrations while /P4's of *C. gratus* have 8-9 serrations with most having nine.

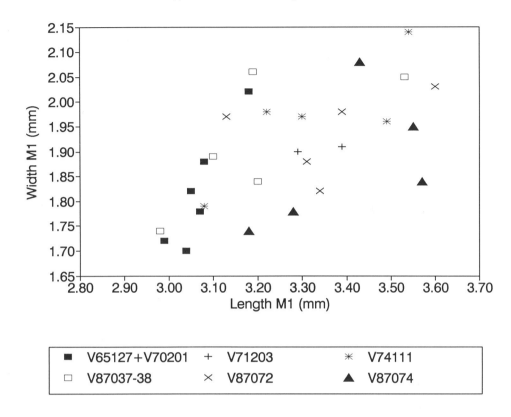

FIGURE 22. Bivariate plot (length vs. width) of M1/'s of *Cimexomys* from Bug Creek Anthills (V65127 and V70201), Harbicht Hill (V71203), Up-Up-the-Creek 2 and 3 (V87037-38), Tedrow Quarry D (V87072) (all from the Hell Creek Formation, McCone County, Montana), and Worm Coulee 1 (V74111; from the Tullock Formation, Garfield County, Montana).

Subclass THERIA Parker and Haswell, 1897

Infraclass TRIBOSPHENIDA McKenna, 1975

Supercohort MARSUPIALIA (Illiger, 1811) Cuvier 1817

Cohort ALPHADELPHIA Marshall, Case, and Woodburne, 1990

Order PERADECTIA Marshall, Case, and Woodburne, 1990

Superfamily PERADECTOIDEA (Crochet, 1979) Marshall, Case, and Woodburne, 1990

Family PERADECTIDAE (Crochet, 1979) Marshall, Case, and Woodburne, 1990

Subfamily CAROLOAMEGINIIDAE Ameghino, 1901

*Glasbius* Clemens, 1966

*Glasbius twitchelli* Archibald, 1982

Tables 23-24

*Glasbius twitchelli* Archibald, 1982, p. 137

   *Type*. UCMP 115853, an isolated right M3/.

   *Type locality*. UCMP loc. V73087, Hell Creek Formation, Montana.

   *Referred specimens*. Two /Mx fragments, 1 /M2, and 1 Mx/ fragment from loc.

Table 23. Measurements of isolated and associated upper molars of *Glasbius twitchelli* from McGuire Creek localities

| Locality | Specimen | Tooth Site | Length | W-A | W-P |
|---|---|---|---|---|---|
| V87037 | 132630 | M2/ | 2.17 | 2.20 | 2.31 |
| V87074 | 133751 | M2/ | 2.08 | 2.27 | 2.42 |
| V87072 | 134710[a] | M2/ | 2.18 | 2.42 | 2.53 |
|  | 134710[a] | M3/ | 1.93 | 2.35 | 2.22 |
| V87072 | 134782 | M3/ | 1.78 | 2.14 | 1.99 |

a: Associated teeth from a maxilla fragment.

V87035. Two /Mx's, 1 M2/, and 1 Mx/ fragment from loc. V87037. Dentary fragment with /P4, /M3-4, UCMP 133069, and 2 /M2's from loc. V87038. Maxillary fragment with M2-3/ UCMP 1347101, 1 /M3, and 1 M3/ from loc. V87072. One M2/, 1 /Mx, and 1 Mx/ from loc. V87074.

*Localities.* UCMP locs. V87035, V87037, V87038, V87072, and V87074.

*Distribution.* Hell Creek Formation, Montana (Lancian and *Puercan); Frenchman Formation, Saskatchewan (Lancian, in part).

*Comments.* Using fossils from the Lance Formation, Wyoming, and a mandibular fragment from the Hell Creek Formation, Montana, Clemens (1966) described a new monogeneric subfamily of didelphids, the Glasbiinae, composed of one species, *Glasbius intricatus*. Archibald (1982), when analyzing mammal faunas from the Hell Creek Formation in Montana, noted several minor morphological characters that could be used to distinguish a new species of *Glasbius*. On the basis of these characters, Archibald (1982) erected *Glasbius twitchelli* and referred the entire Hell Creek Formation sample to it. As a result, *G. intricatus* was restricted to the Lance Formation, Wyoming.

The minor morphological features used to distinguish *G. twitchelli* from *G. intricatus* are also consistently present on fossils from McGuire Creek and clearly support Archibald's (1982) decision to erect a new species for the Hell Creek sample. Upper and lower molars from McGuire Creek referred to *G. twitchelli* are larger than *G. intricatus* (compare Tables 23, 24; with Clemens, 1966, table 6). Also, in all specimens from McGuire Creek, M2/ and M3/'s lack C cusps, M3's only possess a single D cusp, and /M2's and /M3's consistently have one (/M2) or two (/M3) well-developed basal cingulum cusps. These are the main characters used to distinguish *G. twitchelli* from *G. intricatus* (Archibald, 1982).

*Description.* All specimens of *Glasbius* from McGuire Creek are referred to *G. twitchelli*. The McGuire Creek sample is similar in morphology and size (Tables 23, 24) to fossils previously described by Archibald (1982) from Montana. The only difference of note is that /M3's from McGuire Creek are slightly wider (Table 24).

Table 24. Measurements of associated and isolated lower molars and premolars of *Glasbius twitchelli* from McGuire Creek localities

| Locality | Specimen | Tooth Site | Length | W-Tri | W-Tal |
|----------|----------|-----------|--------|-------|-------|
| V87038 | 133070 | /M2 | 1.98 | 1.41 | 1.76 |
| V87038 | 133071 | /M2 | 2.29 | 1.39 | 1.70 |
| V87037 | 132631 | /M2 | 2.18 | 1.25 | 1.51 |
| V87035 | 132521 | /M3 | 2.10 | 1.60 | 1.75 |
| V87072 | 134781 | /M3 | 2.10 | 1.78 | 2.02 |
| V87038 | 133069[a] | /M3 | 2.10 | 1.70 | 1.97 |
|  | 133069[a] | /M4 | 1.55 | 1.21 | 1.06 |
|  | 133069[a] | /P3 | 1.53 | 0.97 | — |
| V87074 | 133752 | /Mx | — | — | 1.91 |
| V87035 | 132522 | /Mx | — | — | 1.83 |
| V87035 | 132520 | /Mx | — | — | 1.83 |
| V87037 | 132629 | /Mx | — | — | 1.91 |

a: Associated teeth from a dentary.

Subfamily ALPHADONTINAE
Marshall, Case, and Woodburne, 1990
*Alphadon* Simpson, 1927b

*Comment.* Systematic studies of *Alphadon* available in 1990 led to the recognition of *A. rhaister*, *A. marshi*, and *A. wilsoni* as present in the upper Hell Creek Formation at McGuire Creek. A fourth, unnamed species of *Alphadon*, intermediate in size between *A. marshi* and *A. wilsoni*, briefly described by Johnston and Fox (1984) and Fox (1989) from the Long Fall (Ravenscrag Formation) and Wounded Knee (Frenchman Formation) local faunas in Canada, was also present at McGuire Creek. Recently, Storer (1991) named a new species of *Alphadon*, *A. jasoni*, from the Gryde Local Fauna (Late Cretaceous), Saskatchewan, and referred the entire *Alphadon* sample from Wounded Knee to it. Also, Storer (1991) referred an upper molar from Bug Creek Anthills to *A. jasoni*, but did not mention if the *Alphadon* sample from the Long Fall locality contained *A. jasoni*.

More recently, Zerina Johanson (1991) presented a preliminary revision of *Alphadon marshi* and *Alphadon wilsoni* in which two upper molar morphologies are recognized, based on various characters including shape of the centrocrista and position of stylar cusp C. The first is similar to *A. marshi* and includes the type of *A. wilsoni* (*A. wilsoni* becomes the junior synonym of *A. marshi*) and the other is similar to the morphology of *A. jasoni* described by Storer (1991). Judd Case is also working on a major systematic revision of *Alphadon* (pers. comm., 1991). Therefore, specimens from McGuire Creek presently referred to *A. wilsoni* or *Alphadon* species indeterminate may be eventually referred to *A. jasoni* or *A. marshi*. Until final revisions by Johanson and Case are available, specimens from McGuire Creek will be referred to *A. wilsoni* and

*Alphadon* species indeterminate, but the species *A. "wilsoni"* will be recognized in quotations pending its formalized synonymy with *A. marshi*.

*Alphadon "wilsoni"* Lillegraven, 1969
Tables 25-26

*Alphadon wilsoni* Lillegraven, 1969, p. 39.

*Type.* UA 3681, fragmentary right maxilla with M2-3/ (Lillegraven, 1969, fig. 19-4).

*Type locality.* Loc. KUA-1, Scollard Formation, Alberta.

*Referred specimens.* One M1/ from UCMP loc. V87037. Two M1/'s, 1 M2/, 1 /Mx, from UCMP loc. V87072. One /Mx from UCMP loc. V87073. Two /Mx's, 1 /M4 from UCMP loc. V87074. One /M1 from UCMP loc. V87101.

*Localities.* UCMP locs. V87037, V87072, V87073, V87074, and V87101.

*Distribution.* Scollard Formation, Alberta, and Lance Formation, Wyoming (both Lancian); Hell Creek Formation, Montana (Lancian and *Puercan); possibly Frenchman Formation, Saskatchewan (Lancian, in part).

*Description.* All upper molars referred to *A. "wilsoni"* from McGuire Creek have a large stylar C cusp that is well separated from cusp B. This development of the stylar cusps distinguishes *A. "wilsoni"* from *Protalphadon lulli*, which has a small stylar cusp C that is not well separated from cusp B (Clemens, 1966; Cifelli, 1990). In size and proportions, the McGuire Creek fossils are similar to those described by Lillegraven (1969). Many of these specimens may eventually be referred to *A. jasoni* or *A. marshi* (see discussion above). Dental measurements are given in Table 25.

Isolated lower molars of smaller species of *Alphadon* are difficult to separate with confidence and are distinguished by differences in size and proportions (Clemens, 1966, 1973; Lillegraven, 1969), but there is some degree of overlap in their ranges of variation. Specimens from McGuire Creek referred to *A."wilsoni"* are similar in size and proportions to the teeth described by Lillegraven (1969) and can be separated from *A. marshi* by these criteria (but see discussion above). Dental measurements are given in Table 26.

---

Table 25. Measurements of isolated upper molars of *Alphadon "wilsoni"* from McGuire Creek localities

| Locality | Specimen | Tooth Site | Length | W-A | W-P | L/ W-A | L/ W-P |
|---|---|---|---|---|---|---|---|
| V87072 | 134778 | M1/ | 2.06 | 2.07 | 2.27 | 1.00 | .91 |
| V87072 | 134709 | M1/ | 2.05 | 2.01 | 2.11 | 1.02 | .97 |
| V87037 | 132625 | M1/ | 1.96 | 1.90 | 2.28 | 1.03 | .86 |
| V87072 | 134776 | M2/ | 2.49 | 2.52 | 2.60 | .99 | .96 |

Table 26. Measurements of isolated lower molars of *Alphadon "wilsoni"* from McGuire Creek localities

| Locality | Specimen | Tooth Site | Length | W-Tr | W-Ta | L/ W-Tr | L/ W-Ta |
|----------|----------|------------|--------|------|------|---------|---------|
| V87101 | 133822 | /M1 | 1.66 | .77 | .85 | 2.16 | 1.95 |
| V87073 | 133522 | /M2-3? | 2.26 | 1.37 | 1.34 | 1.65 | 1.69 |
| V87074 | 133755 | /M2-3? | 2.26 | 1.41 | 1.36 | 1.60 | 1.66 |
| V87074 | 133756 | /M2-3? | 2.13 | 1.38 | 1.37 | 1.54 | 1.55 |
| V87072 | 133382 | /M2-3? | 2.35 | 1.31 | 1.33 | 1.79 | 1.77 |
| V87074 | 133757 | /M4 | 1.99 | 1.37 | 1.10 | 1.45 | 1.81 |

*Alphadon marshi* Simpson, 1927b

*Alphadon marshi* Simpson, 1927b, p. 125

   *Type.* YPM 13659, right M3/ (Simpson, 1929a, fig. 45F).

   *Type locality.* Lance Formation, Niobrara County, Wyoming.

   *Referred specimen.* One /M4, UCMP 134777, from loc. V87072.

   *Locality.* UCMP loc. V87072.

   *Distribution.* Possibly Fruitland and Kirtland formations, New Mexico (Edmontonian); Lance Formation, Wyoming, and Scollard Formation, Alberta (both Lancian); Hell Creek Formation, Montana (Lancian and *Puercan).

   *Revised diagnosis.* Clemens, 1966, p. 5, and Lillegraven, 1969, p. 33.

   *Description.* A single specimen is referred to *A. marshi* on the basis of size and the proportions of length to trigonid and talonid width as described by Lillegraven (1969). Measurements of UCMP 134777 are: L: 2.38; W/Tri: 1.47; W/Tal: 1.27; L/W-Tri: 1.62; L/W-Tal: 1.87.

*Alphadon rhaister* Clemens, 1966

Table 27

*Alphadon rhaister* Clemens, 1966, p. 11.

   *Type.* UCMP 50292, fragmentary left maxilla with M1-3/.

   *Type locality.* UCMP loc. V5815, Lance Formation, Wyoming.

   *Referred specimens.* One /Mx talonid from loc. V87036. One /Mx trigonid from loc. V87052. One /M2 or /M3, from loc. V87072. One /Mx from loc. V87074. Two /M1's, from loc. V87098.

   *Localities.* UCMP locs. V87036, V87052, V87072, V87074, and V87098.

   *Distribution.* Possibly St. Mary River Formation, Alberta (Edmontonian); Lance Formation, Wyoming, and Scollard Formation, Alberta (both Lancian); Hell Creek Formation, Montana (Lancian and *Puercan).

   *Description.* Fossils from McGuire Creek referred to *Alphadon rhaister* are similar to those described by Clemens (1966) from the type Lance Formation. The positioning of the intersection of the trigonid and the cristid obliqua below the protoconid-metaconid

Table 27.  Measurements of isolated lower molars of *Alphadon rhaister* from McGuire Creek localities

| Locality | Specimen | Tooth Site | Length | W-Tri | W-Tal |
|----------|----------|------------|--------|-------|-------|
| V87098 | 134651 | /M1 | 3.23 | 1.73 | 1.94 |
| V87098 | 134648 | /M1 | 3.29 | 1.65 | 1.86 |
| V87074 | 133758 | /M1 | 3.14 | — | — |
| V87072 | 134708 | /M2 or /M3 | 3.65 | 1.96 | 2.08 |
| V87036 | 132491 | /Mx | — | — | 2.07 |
| V87052 | 132357 | /Mx | — | 2.27 | — |

notch separates them from teeth of similar-sized species of *Pediomys*. In size (Table 27) they are consistently within the range of variation exhibited by *Alphadon rhaister* (Clemens, 1966, table 19). Morphologically, they add no new information to the original description of the species.

*Alphadon* sp indet.
Table 28

*Referred specimens.* One M2/ from UCMP loc. V87035. One /M1 from UCMP loc. V87037. One M4/ from UCMP loc. V87072. One /M1 from UCMP loc. V87098. Five fragmentary molars from various UCMP localities.

*Localities.* UCMP locs. V87035, V87037, V87038, V87072, V87074, and V87098.

*Description.* The four specimens listed in Table 28 are intermediate in size and proportions between those of *Alphadon marshi* and *A. "wilsoni"*. Recently, Johnston and Fox (1984) and Fox (1989) described isolated molars of *Alphadon* from the Long Fall (Ravenscrag Formation) and Wounded Knee (Frenchman Formation) localities in Saskatchewan which are also intermediate in size between *A. "wilsoni"* and *A. marshi*. Apparently these fossils belonged to an as then undescribed species of the genus (referred to as *Alphadon* sp.) which was to be named and characterized when a larger sample was collected (Johnston and Fox, 1984; Fox, 1989). Subsequently, Storer (1991) referred all *Alphadon* specimens from Wounded Knee to a new species, *A. jasoni*.

McGuire Creek fossils of *Alphadon* within the size range that Johnston and Fox (1984) and Fox (1989) assigned to the unnamed species may be conspecific with *A. jasoni* or *A. marshi*. Pending formal publication of the revision of species of *Alphadon* in progress by Case and Johanson (described above), the four McGuire Creek specimens listed in Table 28 are referred to *Alphadon* sp. indeterminate.

The other five specimens included here are small but are too fragmentary to be referable to any of the small species of *Alphadon* with confidence.

Table 28. Measurements of isolated upper and lower molars of *Alphadon* sp. indet. from McGuire Creek localities

| Locality | Specimen | Tooth Site | Length | W-A | W-P | L/ W-A | L/ W-P |
|----------|----------|-----------|--------|-----|-----|--------|--------|
| V87035 | 132518 | M2/ | 2.74 | 2.68 | 2.87 | 1.02 | .95 |
| V87072 | 133463 | M4/ | 1.76 | 2.25 | 1.91 | .78 | .92 |

| Locality | Specimen | Tooth Site | Length | W-Tr | W-Ta | L/ W-Tr | L/ W-Ta |
|----------|----------|-----------|--------|------|------|---------|---------|
| V87098 | 134653 | /M1 | 2.06 | 1.20 | 1.23 | 1.72 | 1.67 |
| V87037 | 132627 | /M1 | 2.25 | 1.30 | 1.38 | 1.73 | 1.63 |

Subfamily PERADECTINAE (Crochet, 1979) Reig, 1981
*Peradectes* Matthew and Granger, 1921
*Peradectes* cf. *P. pusillus* Matthew and Granger, 1921
Table 29

*Thylacodon pusillus* Matthew and Granger, 1921, p. 2.

    *Type.* AMNH 16414, dentary fragment with M2-3, M1 talonid (Matthew, 1937, fig. 82).

    *Type locality.* "...upper level of the Puerco Formation near Ojo Alamo, San Juan Basin, New Mexico..." (Matthew and Granger, 1921, p. 2).

    *Referred specimens.* Associated M1/, UCMP 132299, and M2/, UCMP 132300, from loc. V84194. One /M1 from loc. V87035. One /M2 from loc. V87038. One M3/, 1 /M2 from loc. V87037. One M1/, 1 /M3 from loc. V87072. One M1/ from loc. V87074.

    *Localities.* UCMP locs. V84194, V87035, V87037, V87038, V87072, and V87074.

    *Distribution.* Upper Hell Creek and lower Tullock formations, Montana (both Puercan).

    *Comments.* Reference of this material to *Peradectes* cf. *P. pusillus* is based on comparison with material from the Hell's Hollow Local Fauna (UCMP locs. V74110 and V74111) described by Archibald (1982). Archibald's (1982) referral of the Hell's Hollow material to *Peradectes* cf. *P. pusillus* was based on an unpublished revision of North American didelphids in progress by W. A. Clemens at that time, which included the referral of *Thylacodon pusillus* to the genus *Peradectes.* Alternatively, Krishtalka and Stucky (1983) have argued that *Thylacodon pusillus* may not be a species of *Peradectes.* Clemens has now included the Hell's Hollow material in his study and the manuscript is presently in the final stages of preparation (pers. comm., Clemens, 1989), publication of which should aid in settling this question.

    *Description.* *Peradectes* cf. *P. pusillus* was adequately described by Archibald (1982). With the exception of one specimen, all teeth referred to this species from McGuire Creek closely resemble those from Hell's Hollow in size and morphology (Table 29). An isolated M1/, UCMP 134779 from V87072, differs in the morphology of the C region of the stylar shelf. All M1/'s from Hell's Hollow have a large B cusp and slightly smaller

Table 29. Measurements of isolated and associated upper and lower molars of
*Peradectes* cf. *P. pusillus* from McGuire Creek localities

| Locality | UCMP Specimen | Tooth Site | Length | W-A/Tri | W-P/Tal |
|----------|---------------|------------|--------|---------|---------|
| V84194   | 132299[a]     | M1/        | 2.43   | 2.17    | 2.44    |
|          | 132300[a]     | M2/        | 2.40   | 2.80    | 2.92    |
| V87074   | 133570        | M1/        | 2.20   | 2.18    | 2.38    |
| V87072   | 134779        | M1/        | 2.28   | 2.03    | 2.27    |
| V87037   | 132624        | M3/        | 2.19   | 2.99    | 3.24    |
| V87035   | 132519        | /M1        | 2.39   | 1.20    | 1.27    |
| V87038   | 133072        | /M2        | 2.51   | 1.34    | 1.53    |
| V87072   | 134780        | /M2        | 2.49   | 1.39    | 1.35    |
| V87037   | 134684        | /M2        | 2.41   | 1.51    | 1.38    |

a: Associated teeth probably from a single maxilla fragment.

and subequal C and D cusps (Archibald, 1982). These 3 stylar cusps are well separated.
UCMP 134779 has 2 stylar cusps in the region normally occupied by a single C cusp in
the Hell's Hollow specimens. The anterior of the two is very small and is not well separated from cusp B. It is situated on the labial edge of the stylar shelf anterior to a slight
indentation on the midline of the shelf, which is suggestive of a weakly developed
ectoflexus. The posterior C cusp is larger, but still much smaller and well separated
from cusp D and the other more anteriorly positioned C cusp. The posterior C cusp is
shortened in an antero-posterior orientation, giving the cusp a sharp ridge-like appearance. Otherwise, UCMP 134779 is very similar to M1/'s from Hell's Hollow.

Order SPRASSODONTA (Ameghino, 1894) Marshall, Case, and Woodburne, 1990
Superfamily BORHYAENOIDEA (Ameghino, 1894) Simpson, 1930
Family STAGODONTIDAE Marsh, 1889b
*Didelphodon* Marsh, 1889a
*Didelphodon vorax* Marsh, 1889a
Table 30

*Didelphodon vorax* Marsh, 1889a, p. 88 (See Clemens, 1966, p. 60, for synonymies).
   *Type.* YPM 11827, left M2/ (Marsh, 1889a, pl. IV, figs. 1-3).
   *Type locality.* Mammal locality no. 1 of Lull (1915), UCMP loc. V5003, Lance
Formation, Wyoming.
   *Referred specimens.* Left maxillary fragment with complete or partial alveoli for C/,
and P1-3/, UCMP 134795; left edentulous dentary fragment, UCMP 134793; maxillary
fragment with labial alveoli of Mx, UCMP 134794; 1 /M4 fragment, 1 /P1 fragment, 1
/P1, 1 Mx/ fragment, 1 P3/, from loc. V85092. One P2/ from loc. V87029. One /P3, 1
M1/ from loc. V87034. One /Px fragment, 1 /M4 talonid from loc. V87035. One P3/

Table 30. Measurements of isolated molars and premolars of *Didelphodon vorax* from McGuire Creek localities[a]

| Locality | Specimen | Tooth Site | Length | Width | |
|---|---|---|---|---|---|
| | | | | W-A/Tri | W-A/Tal |
| V87029 | 132339 | P2/ | 5.74 | 6.09 | |
| V85092 | 132824 | P3/ | 6.42 | 6.75 | |
| V87034 | 132468 | M1/ | 4.37[b] | 5.64 | 6.51[b] |
| V85092 | 134563 | /P1 | 4.70 | 5.56 | |
| V87072 | 133516 | /P2 | 5.71 | 4.97 | |
| V87034 | 132467 | /P3 | 6.04 | 4.74 | |
| V87072 | 133383 | /DP3 | 3.76 | 2.53 | |
| V87070 | 133156 | /M2 | 5.36[b] | 4.35[b] | — |
| V87035 | 132512 | /Mx | — | — | 4.82[b] |
| V85092 | 132826 | /M4 | — | 5.21[b] | — |

a. All sites are Puercan except V85092 which is Lancian.
b: Approximate measurement.

fragment from loc. V87037. One /M2 from loc. V87070. One /DP3 from loc. V87072. One Mx/ fragment from loc. V87073. One Mx/ fragment from loc. V87078. Left edentulous dentary fragment, UCMP 132807, from loc. V87083. One DP3/ fragment from loc. V87098. One /Mx fragment from loc. V87151.

*Localities.* UCMP locs. V85092, V87029, V87034, V87035, V87037, V87070, V87072, V87073, V87078, V87083, V87098, and V87151.

*Distribution.* Lance Formation, Wyoming, Scollard Formation, Alberta, and Hell Creek Formation, South Dakota (all Lancian); Hell Creek Formation, Montana (Lancian and *Puercan).

*Revised diagnosis.* Clemens, 1966, p. 60, with modifications by Clemens, 1968, pp. 4-10.

*Description.* Isolated teeth from McGuire Creek referred to *Didelphodon vorax* are similar to those from the Lance Formation, Wyoming, described by Clemens (1966), with one exception. UCMP 134563, a /P1, is larger (Table 30) than those from the Lance Formation (Clemens, 1966, table 13), but with regard to the range in variation of premolar size exhibited by the species this is not considered significant. Dental measurements are given in Table 30.

*Discussion.* The dental arcade of stagodontid marsupials is notable for the presence of bulbous premolars, each with a massive main cusp and a large accessory lobe. Configuration of the upper premolars of *Didelphodon* and other stagodontid marsupials was uncertain because of the lack of specimens preserving this portion of the skull. Isolated upper premolars of *Didelphodon* were interpreted to be oriented with their accessory lobes on either their labial (Clemens, 1966) or lingual (Clemens, 1968; Lillegraven, 1969; Archibald, 1982; Fox and Naylor, 1986) sides.

Lofgren (1992) presented the upper premolar orientation of *Didelphodon vorax* using an edentulous maxillary fragment (UCMP 134795) from the upper Hell Creek Formation at McGuire Creek (UCMP loc. V85092). The maxillary fragment contains a number of complete or partial alveoli, into which isolated premolars of *D. vorax* from the type Lance Formation, Wyoming, were placed. This process indicates that the accessory lobes are lingual on P2/-P3/ and only a single alveolus is available for P1/ (Lofgren, 1992). Isolated premolars identified as P1/'s in previous studies have two roots (Clemens, 1966; Lillegraven, 1969), but these teeth may actually be P2/'s (Lofgren, 1992). In any case, the only definitive statement concerning P1/ morphology is provided by UCMP 134795, which has but a single alveolus available for P1/.

Cohort AUSTRALIDELPHIA Szalay, 1982
Order MICROBIOTHERIA Ameghino, 1889
Superfamily MICROBIOTHERIODEA (Ameghino, 1887) Reig, Kirsch, and
Marshall, 1985
Family PEDIOMYIDAE (Simpson, 1927a) Clemens, 1966
*Pediomys* Marsh, 1889a

A systematic revision of *Pediomys* is presently underway by R. Cifelli. These revisions are not available at present, but apparently changes in generic reference are forthcoming.

*Pediomys krejcii* Clemens, 1966
Table 31

*Pediomys krejcii* Clemens, 1966, p. 43.

   *Type.* UCMP 51390, fragmentary left maxilla with P3/, M1-3/ (Clemens, 1966, fig. 26).

   *Type locality.* UCMP loc. V5711, Lance Formation, Wyoming.

   *Referred specimens.* Right dentary fragment with M4, UCMP
132490, from loc. V87036.  Two Mx/'s, 1/Mx, from loc. V87037.  One /M2 from loc.

---

Table 31. Measurements of isolated upper and lower molars of *Pediomys krejcii* from McGuire Creek localities

| Locality | Specimen | Tooth Site | Length | W-A/Tri | W-P/Tal |
|---|---|---|---|---|---|
| V87151 | 132176 | M1/ | 1.54 | 1.41 | 1.80 |
| V87037 | 132621 | Mx/ | — | — | 1.41 |
| V87037 | 132622 | Mx/ | — | — | 2.17 |
| V87072 | 133461 | /M2 | 1.47 | .93 | .96 |
| V87074 | 133759 | /M3 | 1.62 | 1.03 | 1.03 |
| V87036 | 132490 | /M4 | 1.63 | .94 | .80 |
| V87037 | 134683 | /Mx | — | — | .81 |
| V87151 | 134685 | /M1 | — | .79 | — |

V87072. One /M3 from loc. V87074. One M1/, 1 /M1 trigonid, both from loc. V87151.

*Localities.* UCMP locs. V87036, V87037, V87072, V87074, and V87151.

*Distribution.* Possibly St. Mary River Formation, Alberta (Edmontonian); Lance Formation, Wyoming, and Scollard Formation, Alberta (both Lancian); Hell Creek Formation, Montana (Lancian and *Puercan).

*Description.* Reference of the upper molars listed in Table 31 to *Pediomys krejcii* is based on their small size and the absence of a stylar cusp C on UCMP 132176. *P. krejcii* is distinguished from *P. cooki* by its small size, and can be separated from *P. elegans* by the absence of stylar cusp C. UCMP 132176 differs slightly from the original description of the species given by Clemens (1966) in that it has a distinct anterolingual cingulum. Lillegraven (1969) noted the presence of this character in the *P. krejcii* sample from the upper Edmonton Formation of Alberta, but considered it to have little taxonomic significance because of the highly variable development of the cingula.

Specimens of the lower dentition from McGuire Creek referred to *P. krejcii* are similar to those from the type Lance Formation described by Clemens (1966). Dental measurements are presented in Table 31.

*Pediomys elegans* Marsh, 1889a
Tables 32-33

*Pediomys elegans* Marsh, 1889a, p. 89.

*Type.* YPM 11866, broken and heavily weathered upper molar (Marsh, 1889a, pl. IV, figs. 23-25).

*Neotype.* CM 11658, fragmentary right maxilla with M2-4, designated as neotype by Simpson (1929b, p. 111).

*Type locality.* Mammal locality no. 1 of Lull (1915) = UCMP loc. V5003. Neotype from somewhere in the type Lance Formation, probably loc. V5003.

*Referred specimens.* One M2/ from loc. V87028. One M1/ from loc. V87037. One /M1, 1 /M2, from loc. V87038. One /M3, 1 M3/, from loc. V87051. One /M4 from loc.

Table 32. Measurements of isolated upper molars of *Pediomys elegans* from McGuire Creek localities

| Locality | Specimen | Tooth Site | Length | W-A | W-P |
|----------|----------|------------|--------|------|------|
| V87072 | 133377 | M1/ | 2.01 | 1.80 | 2.26 |
| V87072 | 133375 | M1/ | 2.00 | 1.87 | 2.41 |
| V87037 | 132623 | M1/ | — | 1.74 | — |
| V87098 | 133854 | M2/ | 2.05 | 2.48 | 3.00 |
| V87028 | 134624 | M2/ | — | — | 2.60 |
| V87074 | 133760 | M3/ | 1.86 | 2.17 | 2.71 |
| V87074 | 133761 | M3/ | — | 2.52 | — |
| V87051 | 132418 | M3/ | — | — | 3.16 |

Table 33. Measurements of associated and isolated lower molars of *Pediomys elegans* from McGuire Creek localities

|  | Number | Observed Range | Mean |
|---|---|---|---|
| /M1 |  |  |  |
| Length | 4 | 1.92-1.98 | 1.96 |
| Trigonid Width | 4 | 1.04-1.27 | 1.18 |
| Talonid Width | 4 | 1.17-1.43 | 1.30 |
| /M2 |  |  |  |
| Length | 3 | 2.05-2.10 | 2.07 |
| Trigonid Width | 3 | 1.01-1.33 | 1.18 |
| Talonid Width | 3 | 1.20-1.37 | 1.30 |
| /M3 |  |  |  |
| Length | 4 | 2.14-2.38 | 2.28 |
| Trigonid Width | 4 | 1.51-1.60 | 1.56 |
| Talonid Width | 6 | 1.47-1.78 | 1.66 |
| /M4 |  |  |  |
| Length | 4 | 1.94-2.24 | 2.10 |
| Trigonid Width | 5 | 1.19-1.35 | 1.28 |
| Talonid Width | 4 | 1.15-1.40 | 1.25 |

V87052. Two /M3's, 1 /M4, from loc. V87071. Left dentary fragment with /M1, /M3-4, UCMP 134773, left dentary fragment with /M3-4, UCMP 133462, 2 M1/'s, 2 /M1's, 2 /M2's, 1 /M4, 2 /Mx's, all from loc. V87072. Two M3/'s, 1 /M3, 1 /Mx, from loc. V87074. One M2/ from loc. V87098.

*Localities.* UCMP locs. V87028, V87037, V87038, V87051, V87052, V87071, V87072, V87074, and V87098.

*Distribution.* Lance Formation, Wyoming, and Scollard Formation, Alberta (both Lancian); Frenchman Formation, Saskatchewan (Lancian, in part); Hell Creek Formation, Montana (Lancian and *Puercan); Ravenscrag Formation, Saskatchewan (*Puercan).

*Revised diagnosis.* Clemens, 1966, p. 35.

*Description.* Upper molars referred to *P. elegans* agree closely in size (Table 32) and stylar cusp morphology to Clemens' (1966) description of the species. A large or distinct stylar cusp C is evident on all specimens. This feature separates *P. elegans* from the similar-sized species of *Pediomys*, *P. cooki* (Clemens, 1966). Also, the McGuire Creek fossils tend to have a large D cusp which is usually small in *P. cooki* (Clemens, 1966).

Dental measurements of lower molars of *P. elegans* are presented in Table 33. Lower molars of *P. elegans* and *P. cooki* are distinguished primarily by size (Clemens, 1966). The specimens from McGuire Creek consistently fall into the size range displayed by *P. elegans* (Clemens, 1966, table 22). However, there is some degree of overlap in size between the two species and some fossils referred to *P. elegans* may actually belong to *P. cooki*. Although possible, this is considered unlikely, because all of the upper molars in this size range from McGuire Creek are clearly referable to *P. elegans*, and most of these come from the same localities as the lower molars.

Morphologically, the lower molars of *P. elegans* from McGuire Creek are similar to Clemens' (1966) description of the species, and only one addition is necessary. All /M3's and /M4's and two-thirds of the other molars (90% of total sample, n=21) have a small entoconulid. In most cases, the entoconid is larger and higher than both the entoconulid and the hypoconulid. In size and height, the hypoconulid and entoconulid are subequal and the entoconulid is usually not well separated from the entoconid except in /M3's. A survey of the *P. elegans* sample from the type Lance Formation of Wyoming (UCMP loc. V5620) shows that only 30% of the lower molars (n=40) have an entoconulid, and the majority of these are /M4's. The sample of *P. elegans* from the Edmonton Formation of Alberta is small, but the only lower molar (an /M2) figured by Lillegraven (1969, figure 22) has this cusp. The variation in this character probably is not of taxonomic significance.

*Pediomys hatcheri* (Osborn, 1898) Simpson, 1927b
Table 34

*Protolambda hatcheri* Osborn, 1898, p. 172 (see Clemens, 1966, p. 45, for synonymies).

*Type.* AMNH 2202, fragmentary right maxilla with M3/, designated as a lectotype by Clemens, 1966, p. 45 (Osborn, 1898, fig. 1A, center); AMNH 2203, isolated M2/, designated as a syntype by Clemens, 1966, p. 45.

*Type locality.* "Lance Creek area, Lance Formation, Wyoming" (Clemens, 1966, p. 45).

*Referred specimens.* One /Mx from loc. V87035. Left dentary fragment with /M3-4, UCMP 132599, from loc. V87037. One M2/, 1 /Mx trigonid, both from loc. V87072. One /M2 from loc. V87098. One M2/ from loc. V87151.

*Localities.* UCMP locs. V87035, V87037, V87072, V87098, and V87151.

Table 34. Measurements of isolated and associated upper and lower molars of *Pediomys hatcheri* from McGuire Creek localities

| Locality | Specimen | Tooth Site | Length | W-A/Tri | W-P/Tal |
|---|---|---|---|---|---|
| V87072 | 133381 | M2/ | 3.50 | 3.48 | 4.31 |
| V87151 | 132178 | M2/ | — | 3.66 | — |
| V87035 | 132516 | /Mx | — | 2.30 | — |
| V87072 | 133380 | /Mx | — | 2.45 | — |
| V87098 | 133855 | /M2 | 3.35 | 1.85 | 2.11 |
| V87037 | 132599[a] | /M3 | 3.21 | 2.23 | 2.37 |
| | 132599[a] | /M4 | 4.22 | 2.19 | 2.10 |

a: Associated teeth from a dentary fragment.

*Distribution.* Lance Formation, Wyoming, Scollard Formation, Alberta, and North Horn Formation, Utah (all Lancian); Hell Creek Formation, Montana (Lancian and *Puercan).

*Revised diagnosis.* Clemens, 1966, p. 45.

*Description.* In both the M2/'s referred to *Pediomys hatcheri*, the stylar shelf is continuous across the paracone, the B cusps are absent, and the C cusps are small. In these respects and in size (Table 34), they agree closely with the original description of the species by Clemens (1966). One small difference is noteworthy; UCMP 132178 has a small anterolingual cingulum.

Lower molars of *Pediomys hatcheri* from McGuire Creek agree with Clemens' (1966) original descriptions, and only one addition is necessary. The dentary fragment, UCMP 132599, has a long narrow talonid on /M4. Consequently, the length of this tooth falls outside the range of variation observed in the sample from the type Lance Formation.

*Pediomys florencae* Clemens, 1966
Table 35

*Pediomys florencae* Clemens, 1966, p. 50.

*Type.* UCMP 51440, fragmentary left maxilla with M2-3 (Clemens, 1966, fig. 31).

*Type locality.* UCMP loc. V5820, Lance Formation, Wyoming.

*Referred specimens.* Right edentulous dentary fragment, UCMP 134792, from loc. V85092. One M1/, 1 M4/, from loc. V87035. One /Mx trigonid from loc. V87038. One /M3 from loc. V87040. One M1/ from loc. V87072. One Mx/ fragment, 1 /M2, both from loc. V87151.

*Localities.* UCMP locs. V85092, V87035, V87038, V87040, V87072, and V87151.

*Distribution.* Lance Formation, Wyoming (Lancian); Hell Creek Formation, Montana (Lancian and *Puercan).

*Description.* *Pediomys florencae* is the largest species of the genus, and the fragmentary specimens UCMP 134792 and UCMP 134789 are referred to this taxon primarily on this basis. A large edentulous dentary fragment, UCMP 134792, is also referred to

Table 35. Measurements of isolated upper and lower molars of *Pediomys florencae* from McGuire Creek localities

| Locality | Specimen | Tooth Site | Length | W-A/Tri | W-P/Tal |
|---|---|---|---|---|---|
| V87035 | 132517 | M1/ | 3.29 | 3.01 | 3.82 |
| V87072 | 134775 | M1/ | 2.82[a] | 3.06 | 3.64 |
| V87035 | 132515 | M4/ | 4.10 | 4.30 | 4.19 |
| V87151 | 132179 | /M2 | 4.13 | 2.45 | 2.53 |
| V87040 | 133153 | /M3 | 5.25 | 3.11 | 3.24 |
| V87038 | 134789 | /Mx | — | 3.24 | — |

a: Approximate measurement.

the species. It compares closely to mandibles of *P. florencae* from UCMP locality V5620 from the type Lance Formation.

Two complete lower molars from McGuire Creek are referred to *P. florencae*. One of these, an /M3 (UCMP 133153), is larger than the single /M3 from the type Lance Formation described by Clemens (1966). Until further sampling proves otherwise, this size difference is attributed to variation within the species.

Of the three upper molars referred to *P. florencae*, two provide some additional morphological data. A well preserved M1/, UCMP 132517, has small C and D stylar cusps, and cusp B is absent. Also, the stylar shelf is not continuous across the paracone. On UCMP 132515, an M4/, the stylar shelf extends to a small indentation posterior to a small C cusp, cusp B is absent, and a large parastyle is present. The metacone is much smaller than the paracone and is near the posterolabial edge of the tooth. A weak anterolingual cingulum and both conules are present. Dental measurements are provided in Table 35.

*Pediomys* sp. indet.

*Referred specimens.* Two /Mx talonids from loc. V87037. One /Mx talonid from loc. V87038. One M2/?, UCMP 133181, from loc. V87071.

*Localities.* UCMP locs. V87037, V87038, and V87071.

*Comments.* On UCMP 133181, the stylar shelf labial to the paracone is present but weakly developed, and cusp B is absent. Unfortunately, the posterolabial quarter of the tooth, including the area of the stylar shelf which would contain cusps C and D (if present), is missing. On the basis of size, this specimen could be referred to *P. cooki* or *P. elegans*, but loss of the posterior part of the stylar shelf prevents reference to one or the other species.

Three lower molar talonids are referred to *Pediomys* sp. indeterminate, because they are too fragmentary for species identification.

Infraclass EUTHERIA (Gill, 1872) Huxley, 1880
Cohort EPITHERIA McKenna, 1975
Superorder INSECTIVORA (Illiger, 1811) Novacek, 1986
Order LEPTICTIDA (McKenna, 1975) Novacek, 1986
Family GYPSONICTOPIDAE
*Gypsonictops* Simpson, 1927a
*Gypsonictops illuminatus* Lillegraven, 1969
Tables 36-37; Figure 23

*Gypsonictops illuminatus* Lillegraven, 1969, p. 51.

*Type.* UA 2447, right maxillary fragment with P3-4/, M1-3/ (Lillegraven, 1969, fig. 27-6).

*Type locality.* Loc. KUA-1, Scollard Formation, Alberta.

*Referred specimens.* One P4/ fragment from loc. V85092. One /Mx, 1 Mx/, from loc. V87035. One /M1, 2 DP4/'s, 1 /DP4, from loc. V87038. One /M2, 1 M3/, from loc. V87051. Two /M1's from loc. V87071. One /P4, 2 /M3's, 1 M1/, 1 Mx/, from loc.

Table 36. Measurements of isolated lower molars and premolars of *Gypsonictops illu-minatus* from McGuire Creek localities

| Locality | Specimen | Tooth Site | Length | W-Tri | W-Tal |
|----------|----------|------------|--------|-------|-------|
| V87072 | 133459 | /P4 | 2.40[a] | 1.48 | 1.26 |
| V87071 | 133233 | /M1 | 2.45 | 1.66 | 1.67 |
| V87038 | 133089 | /M1 | 2.29 | 1.66 | 1.74 |
| V87074 | 133765 | /M1 | 2.31 | 1.67 | 1.66 |
| V87071 | 133232 | /M1 | 2.31 | 1.58 | 1.58 |
| V87074 | 133766 | /M2 | 2.30 | 1.78 | 1.54 |
| V87051 | 132416 | /M2 | 2.31 | 1.59 | 1.53 |
| V87072 | 134765 | /M3 | 2.26 | 1.66 | 1.43 |
| V87072 | 133458 | /M3 | 2.41 | 1.49 | 1.27 |
| V87038 | 133079 | /DP4 | 3.26 | .99 | 1.23 |

a: Approximate measurement.

V87072. One /M1, 1 /M2, from loc. V87074. One /DP4, 1 P3/, from loc. V87098. One M3/ from loc. V87101. 1 M2/, 1 P4/, from loc. V87151. One Mx/ fragment from loc. V87153.

*Localities.* UCMP locs. V85092, V87035, V87038, V87051, V87071, V87072, V87074, V87098, V87101, V87151, and V87153.

*Distribution.* Scollard Formation, Alberta (Lancian); possibly Frenchman Formation, Saskatchewan (Lancian, in part); Hell Creek Formation, Montana (Lancian and *Puercan); Ravenscrag Formation, Saskatchewan (*Puercan).

*Description.* Lower molars from McGuire Creek referred to *Gypsonictops illuminatus* are consistently within the size range of *G. illuminatus* from the Scollard Formation of Alberta described by Lillegraven (1969). The dentition of *G. hypoconus*, a smaller species known from the Lance Formation of Wyoming (Clemens, 1973), is significantly smaller. The other character that differentiates the lower molars of these two species is the relative height of the trigonid: *G. illuminatus* has proportionately lower trigonids than *G. hypoconus* (Lillegraven, 1969; Clemens, 1973). The /M1's and /M2's are identified by the method outlined by Clemens (1973). A single /DP4 (UCMP 133079) is referred to *G. illuminatus*, on the basis of size and comparison to /DP4's of *G. illuminatus* described by Lillegraven (1969, figure 27, 2a-c). Dental measurements are given in Table 36.

Upper molars and premolars from McGuire Creek also consistently fall within the range of variation in size displayed by *G. illuminatus* and are significantly larger than

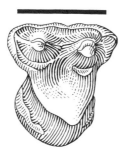

FIGURE 23. *Gypsonictops illuminatus* Lillegraven. Isolated right DP4/, UCMP 133077, locality V87038, occlusal view. Scale bar = 2 mm.

those of *G. hypoconus*. Also, unworn upper molars from McGuire Creek are similar to *G. illuminatus* in the development of an anterior recumbency in the protocone. Three DP4/'s from McGuire Creek are referred to *G. illuminatus*. Previously, DP4/'s of *G. illuminatus* were unknown. Fossils from McGuire Creek are identical to DP4/'s of *G. hypoconus* described by Clemens (1973) from the Lance Formation, but are significantly larger and have an anteriorly recumbent protocone (Figure 23). Dental measurements are given in Table 37.

Table 37. Measurements of isolated upper molars and premolars of *Gypsonictops illuminatus* from McGuire Creek localities

| Locality | Specimen | Tooth Site | Length | W-A | W-P |
|----------|----------|------------|--------|-----|-----|
| V87098 | 134647 | P3/ | 2.50 | Width=2.40 | |
| V87151 | 132172 | P4/ | 2.93 | 3.01 | 3.20 |
| V87038 | 133077 | DP4/ | 2.29 | 2.52 | 2.56 |
| V87098 | 134646 | DP4/ | 2.22 | 2.54 | 2.58 |
| V85092 | 134800 | M1/? | — | 3.65[a] | 3.70[a] |
| V87072 | 133407 | M1/ | 2.88 | 3.55 | 3.57 |
| V87151 | 132173 | M2/ | 2.50 | 3.58 | 3.74 |
| V87051 | 132407 | M3/ | — | — | 1.93 |

a: Approximate measurement.

Infraclass EUTHERIA, incertae sedis
Family PALAEORYCITIDAE (Winge, 1917) Simpson, 1931
*Batodon* Marsh, 1892
*Batodon* tenuis Marsh 1892
Table 38

*Batodon tenuis* Marsh, 1892, p. 258.

*Type.* USNM 2139, fragmentary left dentary with part of the canine, alveolus of /P1, and /P2-4 (Marsh, 1892, pl. XI, fig. 5; Simpson, 1929a, fig. 55; Clemens, 1973, fig. 25d and e).

*Type locality.* Mammal locality no. 1 of Lull (1915); UCMP loc. V5003, Lance Formation, Wyoming.

*Referred specimens.* One /P4, 1 M2/ from loc. V87038. One /M2 from loc. V87074. One /M2 from loc. V87151.

*Localities.* UCMP locs. V87038, V87074, and V87151.

*Distribution.* Lance Formation, Wyoming, and Scollard Formation, Alberta (both Lancian); Hell Creek Formation, Montana (Lancian and *Puercan).

*Revised Diagnosis.* Clemens, 1973, p. 67-68.

*Description.* The sample of *Batodon tenuis* from McGuire Creek is small, consisting of four teeth from three widely scattered localities. *Batodon tenuis* has been adequately described by Lillegraven (1969) and Clemens (1973), and the specimens from McGuire Creek yield little additional morphological information. Dental measurements are provided in Table 38.

Table 38. Measurements of isolated molars and premolars of *Batodon tenuis* from McGuire Creek localites

| Locality | Specimen | Tooth Site | Length | W-A/Tri | W-P/Tal |
|---|---|---|---|---|---|
| V87038 | 133081 | /P4 | 1.08 | .59 | — |
| V87038 | 133080 | M2/ | 1.00[a] | 1.45[a] | 1.67[a] |
| V87074 | 132174 | /M2 | 1.22 | .93 | — |
| V87151 | 133764 | /M2 | 1.20 | .76 | .71 |

a: Approximate measurement.

*Procerberus* Sloan and Van Valen, 1965
*Procerberus formicarum* Sloan and Van Valen, 1965
Table 39

*Procerberus formicarum* Sloan and Van Valen, 1965, p. 225.

*Type.* UMVP 1460, dentary fragment with /P3-4, /M1 (not figured, but isolated teeth figured, Sloan and Van Valen, 1965, fig. 6; see also Lillegraven, 1969, fig. 32-1,2,3, and Van Valen, 1969, fig. 3).

*Type locality.* Bug Creek Anthills, Hell Creek Formation, Montana.

*Referred specimens.* One M3/ from loc. V84193. One M1/ from loc. V84194. One /M2, 1 M2/ from loc. V87028. Two /P4's from loc. V87030. One /M1, 1 /M2, 2 /M3's, 1 P3/, 1 P4/ from loc. V87035. One /M1 from loc. V87036. Two /P4's, 1 /M1, 1 /M2, 1 /M3, 1 P3/, 2 P4/'s, 2 M1/'s, 2 M2/'s, 1 M3/, 2 Mx/'s from loc. V87037. Four /P4's, 3 /M1's, 4 /M2's, 4 /M3's, 1 DP4?/, 2 P4/'s, 2 M1/'s, 3 M2/'s, 2 M3/'s from loc. V87038. One /P4, 2 /M1's, 1 P3/, 1 M1/ from loc. V87051. One /P4, 1 M3/ from loc. V87071. One /P3, 3 /P4's, 6 /M1's, 2 /M2's, 2 /M3's, 1 DP4/, 3 P3/'s, 4 P4/'s, 10 M2/'s, 2 M3/'s from loc. V87072. Two M1/s from loc. V87073. Four /M1's, 1 /M2, 1 /M3, 1 P3/, 5 P4/'s, 2 M1/'s, 2 M2/'s, 2 M3/'s, 3 Mx/'s from loc. V87074. Two /P4's, 1 /M1, 1 /M2, 1 /M3, 1 P4, 2 M1/'s, 1 M2/, 1 Mx/ from loc. V87098. One P4/ from loc. V87114. One /DP4 trigonid, 3 /M1's, 2 /M2's, 4 /M3's, 2 P4/'s, 2 M1/'s, 6 M2/'s, 1 M3/, 1 Mx/ from loc. V87151. One /M3 from loc. V87153.

*Localities.* UCMP locs. V84193, V84194, V87028, V87030, V87035, V87036, V87037, V87038, V87051, V87071, V87072, V87073, V87074, V87098, V87114, V87151, and V87153.

*Distribution.* Upper Hell Creek and lower Tullock Formations, Montana (both Puercan); possibly Ravenscrag Formation, Saskatchewan (Puercan).

*Description. Procerberus formicarum* from the upper Hell Creek Formation at McGuire Creek is similar in size and morphology to fossils from Bug Creek Anthills, the type locality (Table 39). The dental morphology of *P. formicarum* has been described in detail by Sloan and Van Valen (1965) and Lillegraven (1969), and additional comments will be restricted to the DP4/, which is not described in those works.

The only DP4/ (UCMP 134706) that can be confidently referred to *P. formicarum* is unworn. The crown is nearly equal in length and width (L: 2.56; W-A: 2.56; W-P: 2.61), giving the tooth a triangular shape. In gross morphology, the DP4/ of *P. formicarum* resembles those referred to *Cimolestes* by Lillegraven (1969) and Clemens (1973) but differs significantly in the following ways: The paracone and metacone are less conate, and the metacrista and paracrista are less distinct, the stylar shelf is more reduced, with the stylar area labial to the paracone markedly so, the conules are less distinct and lack internal wings, and short and weakly developed lingual cingula are present.

Superorder UNGULATA (Linnaeus, 1766) Novacek, 1986
Order "CONDYLARTHRA"
(Quotation marks indicate a paraphyletic taxon; Prothero et al., 1988.)

Most of the mammalian taxa found in K-T transition local faunas in eastern Montana (upper Hell Creek and lower Tullock formations) have been adequately described in previous works (e.g., Clemens, 1964, 1966, 1973; Sloan and Van Valen, 1965; Lillegraven, 1969; Archibald, 1982). "Condylarths" are a notable exception, because the description of many taxa consists of only one or two sentences given with their original diagnoses (Sloan and Van Valen, 1965; Van Valen, 1978). Archibald (1982) provided more complete descriptions of *Oxyprimus erikseni, Mimatuta morgoth,* and *Mimatuta minuial.* Recently, Luo (1989, 1991) gave detailed descriptions of *Protungulatum donnae, Mimatuta morgoth,* and *Oxyprimus erikseni.* However, *Protungulatum gorgun* and

Table 39. Measurements of isolated molars and premolars of *Procerberus formicarum* from Bug Creek Anthills (V65127) and selected McGuire Creek localities

| | V65127 | V87072 | V87151 | V87037-38 |
|---|---|---|---|---|
| **/P4 Length** | | | | |
| Number | 2 | 2 | - | 4 |
| Observed Range | 2.83-2.88 | 2.60-3.14 | — | 2.67-3.21 |
| Mean | 2.86 | 2.87 | — | 2.89 |
| **/P4 W/Tri** | | | | |
| Number | 2 | 3 | - | 5 |
| Observed Range | 1.20-1.21 | 1.24-1.48 | — | 1.15-1.41 |
| Mean | 1.21 | 1.37 | — | 1.30 |
| **/P4 W/Tal** | | | | |
| Number | 2 | 3 | - | 6 |
| Observed Range | 1.14-1.22 | 1.06-1.33 | — | 1.03-1.33 |
| Mean | 1.18 | 1.22 | — | 1.18 |
| **/M1 Length** | | | | |
| Number | 8 | 5 | 2 | 3 |
| Observed Range | 2.74-2.94 | 2.64-2.88 | 2.61-2.92 | 2.76-2.99 |
| Mean | 2.88 | 2.75 | 2.77 | 2.87 |
| **/M1 W/Tri** | | | | |
| Number | 8 | 5 | 2 | 3 |
| Observed Range | 1.65-1.98 | 1.62-1.77 | 1.70-1.77 | 1.75-1.98 |
| Mean | 1.84 | 1.71 | 1.74 | 1.84 |
| **/M1 W/Tal** | | | | |
| Number | 8 | 5 | 2 | 3 |
| Observed Range | 1.51-1.60 | 1.38-1.56 | 1.52-1.61 | 1.47-1.66 |
| Mean | 1.56 | 1.47 | 1.57 | 1.56 |
| **/M2 Length** | | | | |
| Number | 8 | 2 | 2 | 5 |
| Observed Range | 2.60-2.83 | 2.44-2.71 | 2.68-2.96 | 2.47-3.10 |
| Mean | 2.73 | 2.58 | 2.82 | 2.79 |
| **/M2 W/Tri** | | | | |
| Number | 8 | 1 | 2 | 5 |
| Observed Range | 1.72-2.08 | 1.94 | 1.97-2.03 | 1.63-2.08 |
| Mean | 1.88 | 1.94 | 2.00 | 1.84 |
| **/M2 W/Tal** | | | | |
| Number | 8 | 2 | 2 | 5 |
| Observed Range | 1.47-1.62 | 1.41-1.57 | 1.50-1.65 | 1.30-1.58 |
| Mean | 1.53 | 1.49 | 1.58 | 1.44 |
| **/M3 Length** | | | | |
| Number | 6 | 2 | 4 | 3 |
| Observed Range | 2.91-3.30 | 3.05-3.09 | 2.97-3.24 | 3.10-3.19 |
| Mean | 3.07 | 3.07 | 3.07 | 3.14 |
| **/M3 W/Tri** | | | | |
| Number | 6 | 2 | 4 | 3 |
| Observed Range | 1.69-1.95 | 1.83-1.84 | 1.69-1.96 | 1.92-1.99 |
| Mean | 1.81 | 1.84 | 1.83 | 1.96 |
| **/M3 W/Tal** | | | | |
| Number | 6 | 2 | 4 | 3 |
| Observed Range | 1.41-1.52 | 1.36-1.41 | 1.33-1.52 | 1.44-1.58 |
| Mean | 1.47 | 1.39 | 1.45 | 1.50 |
| **P3/ Length** | | | | |
| Number | 2 | 3 | - | 1 |
| Observed Range | 2.29-2.40 | 2.38-2.81 | — | 2.23 |
| Mean | 2.35 | 2.56 | — | 2.23 |

| | V65127 | V87072 | V87151 | V87037-38 |
|---|---|---|---|---|
| **P3/ Width** | | | | |
| Number | 2 | 3 | - | 1 |
| Observed Range | 2.08-2.33 | 1.93-2.34 | — | 1.49 |
| Mean | 2.21 | 2.10 | — | 1.49 |
| **P4/ Length** | | | | |
| Number | 3 | 1 | 1 | 2 |
| Observed Range | 2.40-2.61 | 2.59 | 2.77 | 2.53-2.60 |
| Mean | 2.51 | 2.59 | 2.77 | 2.57 |
| **P4/ W/A** | | | | |
| Number | 4 | 3 | 1 | 2 |
| Observed Range | 2.37-2.82 | 2.53-2.85 | 2.85 | 2.70-2.79 |
| Mean | 2.60 | 2.63 | 2.85 | 2.75 |
| **P4/ W/P** | | | | |
| Number | 4 | 3 | 2 | 3 |
| Observed Range | 2.69-3.23 | 2.65-2.99 | 2.66-2.93 | 3.00-3.14 |
| Mean | 2.92 | 2.77 | 2.80 | 3.05 |
| **M1/ Length** | | | | |
| Number | 6 | 1 | 3 | 2 |
| Observed Range | 2.43-2.92 | 2.91 | 2.46-2.79 | 2.48-2.67 |
| Mean | 2.66 | 2.91 | 2.62 | 2.58 |
| **M1/ W/A** | | | | |
| Number | 6 | 1 | 3 | 2 |
| Observed Range | 3.17-3.42 | 3.58 | 2.96-3.57 | 3.24-3.38 |
| Mean | 3.30 | 3.58 | 3.29 | 3.31 |
| **M1/ W/P** | | | | |
| Number | 6 | 1 | 3 | 3 |
| Observed Range | 3.32-3.60 | 3.76 | 3.21-3.80 | 3.28-3.43 |
| Mean | 3.46 | 3.76 | 3.51 | 3.34 |
| **M2/ Length** | | | | |
| Number | 2 | 2 | 3 | 2 |
| Observed Range | 2.76-2.76 | 2.65-2.66 | 2.72-2.87 | 2.48-2.70 |
| Mean | 2.76 | 2.66 | 2.78 | 2.59 |
| **M2/ W/A** | | | | |
| Number | 5 | 3 | 3 | 2 |
| Observed Range | 3.08-3.68 | 3.51-3.60 | 3.43-3.80 | 3.64-3.64 |
| Mean | 3.46 | 3.56 | 3.63 | 3.64 |
| **M2/ W/P** | | | | |
| Number | 5 | 5 | 5 | 4 |
| Observed Range | 3.54-4.02 | 3.47-3.89 | 3.54-3.88 | 3.33-3.86 |
| Mean | 3.80 | 3.70 | 3.75 | 3.63 |
| **M3/ Length** | | | | |
| Number | 6 | 2 | 1 | 3 |
| Observed Range | 2.01-2.70 | 2.10-2.32 | 2.31 | 1.94-2.73 |
| Mean | 2.26 | 2.21 | 2.31 | 2.37 |
| **M3/ W/A** | | | | |
| Number | 6 | 2 | 1 | 2 |
| Observed Range | 2.68-3.64 | 2.90-3.12 | 3.61 | 3.56-3.80 |
| Mean | 3.26 | 3.01 | 3.61 | 3.68 |
| **M3/ W/P** | | | | |
| Number | 6 | 2 | 1 | 2 |
| Observed Range | 2.51-3.10 | 2.50-2.66 | 3.07 | 3.01-3.17 |
| Mean | 2.83 | 2.58 | 3.07 | 3.09 |

*Ragnarok nordicum* are still inadequately described and poorly known. A more complete description of the last two species based on new material from McGuire Creek, is given below. *Oxyprimus erikseni, Mimatuta morgoth, Mimatuta minuial*, and *Protungulatum donnae* also are present at McGuire Creek, and additions to their descriptions are given where deemed necessary.

During systematic analysis of the "condylarths" from McGuire Creek, it became apparent that many of the "condylarths" from Mantua Lentil, Polecat Bench Formation (Fort Union Formation), Wyoming, are similar to those from the upper Hell Creek Formation of Montana. Because regional biostratigraphic and biochronological correlations during the Late Cretaceous?-Early Tertiary are based in part on species-level differences within "condylarth" lineages, it became necessary to evaluate the validity of species of genera (*Oxyprimus, Ragnarok*, and *Mimatuta*) that occur in both the Mantua Lentil and Hell Creek areas. Epoxy casts of *Oxyprimus putorius, O. galadrielae, Ragnarok nordicum*, and *Mimatuta minuial* from Mantua Lentil were available for study. Descriptions of these taxa and comparisons to McGuire Creek "condylarths" are given where appropriate.

Family ARCTOCYONIDAE (Giebel, 1855) Murray, 1866
Subfamily OXYCLAENINAE (Scott, 1892) Matthew, 1937
*Protungulatum* Sloan and Van Valen, 1965
*Protungulatum donnae* Sloan and Van Valen, 1965
Tables 40-42, 54

*Protungulatum donnae* Sloan and Van Valen, 1965, p. 226.
    *Type.* SPSM 62-2028, left mandible with /P2-4, /M1-3 (Sloan and Van Valen, 1965, fig. 7).
    *Type locality.* Bug Creek Anthills, Hell Creek Formation, Montana.

---

Table 40. Measurements of associated and isolated upper molars of *Protungulatum donnae* from McGuire Creek localities

| Locality | Specimen | Tooth Site | Length | W-A | W-P | A | A/W-A |
|----------|----------|------------|--------|------|------|------|-------|
| V87034 | 132471 | M1/ | 3.50 | 4.72 | 4.83 | — | — |
| V87036 | 132495 | M1/ | 3.80[a] | 5.08 | 5.30 | 1.90 | .37 |
| V87038 | 133063 | M1/ | 3.75 | 4.70[a] | 4.94[a] | — | — |
| V87101 | 133820 | M1/ | — | — | 5.23 | — | — |
|  | 133820 | M2/ | 4.78 | 5.77 | 6.12 | 1.96 | .34 |
| V87084 | 132811 | M2/ | 3.69 | 5.63 | 5.82 | 2.12 | .38 |
| V87072 | 134772 | M2/ | 4.06 | 5.63 | 5.87 | 2.07 | .38 |
| V87072 | 134696 | M2/ | 4.55 | 6.29 | 6.66 | 2.00 | .32 |
|  | 134696 | M3/ | 4.30 | 5.48 | 4.88 | 1.50 | .27 |
| V87037 | 132614 | M3/ | — | — | 4.64 | 1.78 | — |

a: Approximate measurement.

Table 41. Measurements of isolated and associated lower molars of *Protungulatum donnae* from McGuire Creek localities, and the type of *Protungulatum mckeeveri* from V72210

| Tooth Site | Number | Observed Range | Mean |
|---|---|---|---|
| /M1 | | | |
| Length | 8 | 3.61-4.10 (*4.00) | 3.84 |
| Width-Trigonid | 8 | 2.60-3.15 (*2.65) | 2.81 |
| Width-Talonid | 8 | 2.50-2.96 (*2.70) | 2.71 |
| W-Trigonid/W-Talonid | 8 | 1.00-1.09 (*0.98) | 1.04 |
| Length/Width-Talonid | 8 | 1.36-1.50 (*1.48) | 1.42 |
| /M2 | | | |
| Length | 15 | 3.68-4.52 (*4.40) | 4.19 |
| Width-Trigonid | 15 | 2.89-3.87 (*3.30) | 3.44 |
| Width-Talonid | 15 | 2.46-3.49 (*3.05) | 3.13 |
| W-Trigonid/W-Talonid | 15 | 1.05-1.17 (*1.08) | 1.10 |
| Length/Width-Talonid | 15 | 1.26-1.53 (*1.44) | 1.34 |
| /M3 | | | |
| Length | 7 | 4.42-5.56 (*4.90) | 4.85 |
| Width-Trigonid | 7 | 2.67-3.56 (*3.10) | 3.01 |
| Width-Talonid | 7 | 2.11-2.85 (*2.40) | 2.39 |
| W-Trigonid/W-Talonid | 7 | 1.20-1.29 (*1.30) | 1.26 |
| Length/Width-Talonid | 7 | 1.92-2.12 (*2.04) | 2.03 |

(*): Measurements of the type of *P. mckeeveri* (UCMP 121782).

*Referred specimens.* Dentary with /M2-3, UCMP 132461, from loc. V86031. Dentary with /M2-3, UCMP 132341; 1 /M3 from loc. V87029. Dentary with /M2, UCMP 132440, from loc. V87031. One M1/, 1 /M2 from loc. V87034. Dentary with /M2-3, UCMP 132498; dentary with /M1-2, UCMP 132499; dentary with /M2, UCMP 132497, from loc. V87035. One M1/ from loc. V87036. Dentary with /P4-/M1, UCMP 132595; 2 /M1's, 1 M3/ from loc. V87037. One M1/ from loc. V87038. One /M2 from loc. V87040. Dentary with /M2-3, UCMP 133517; maxilla with M2-3, UCMP 134696; 1 M2/ from loc. V87072. One /M3 from loc. V87082. One M2/ from loc. V87084. Dentary with /M1 talonid, /M2, /M3 trigonid, UCMP 133838; 2 /M1's, 1 /M2, 1 /M3 from loc. V87098. Maxilla with M1-2/, UCMP 133820; 1 /M2 from loc. V87101. One /M2 from loc. V87108. Two /M1's, 1 /M2, 1 M3/ from loc. V87151. One /M2 from loc. V88038.

*Localities.* UCMP locs. V86031, V87029, V87031, V87034, V87035, V87036, V87037, V87038, V87040, V87072, V87082, V87084, V87098, V87101, V87108, V87151, and V88038.

*Distribution.* Upper Hell Creek Formation, Montana (Puercan); possibly lower Tullock Formation, Montana (Puercan); possibly upper Frenchman Formation and Ravenscrag Formation, Saskatchewan (both Puercan).

*Description.* McGuire Creek fossils referable to *Protungulatum* can be separated into

Table 42. Measurements of associated lower molars and premolars of *Protungulatum donnae* from McGuire Creek localities

| Locality | Specimen | Tooth Site | Length | W-Tri | W-Tal |
|----------|----------|-----------|--------|-------|-------|
| V87037 | 132595 | /P4 | 3.51 | 2.27 (width) | |
| | 132595 | /M1 | 3.90 | 2.78 | 2.60 |
| V86031 | 132461 | /M2 | 4.52 | 3.72 | 3.38 |
| | 132461 | /M3 | 5.29 | 3.47 | 2.75 |
| V87029 | 132341 | /M2 | 4.43 | 3.72 | 3.42 |
| | 132341 | /M3 | 5.56 | 3.56 | 2.85 |
| V87035 | 132498 | /M2 | 3.91 | 2.92 | 2.77 |
| | 132498 | /M3 | 4.42 | 2.67 | 2.11 |
| V87098 | 133838 | /M1 | — | — | 2.79 |
| | 133838 | /M2 | 4.41 | 3.55 | 3.27 |
| | 133838 | /M3 | — | 3.28 | — |
| V87035 | 132499 | /M1 | 4.04 | 3.15 | 2.96 |
| | 132499 | /M2 | 4.47 | 3.87 | 3.49 |
| V87072 | 133517 | /M2 | 3.68 | 3.14 | 2.89 |
| | 133517 | /M3 | 4.56 | 2.88 | 2.23 |

two species of similar morphology but of significantly different size ranges. The larger species is the same size as *P. gorgun* from Harbicht Hill, Hell Creek Formation, described by Van Valen (1978). The smaller of the two species is referred to *P. donnae* from Bug Creek Anthills, Hell Creek Formation, described by Sloan and Van Valen (1965).

In addition to the original description, the dentition of *Protungulatum donnae* was described in further detail and contrasted with other Late Cretaceous?-early Paleocene "condylarths" (e.g. *Mimatuta, Oxyprimus,* and *Ragnarok*) by Archibald (1982), Middleton (1983), and Luo (1989, 1991). Therefore, only a few comments are required concerning additional morphological data from McGuire Creek.

Upper molars of *Protungulatum donnae* are similar in size to those of *Mimatuta*, but can be separated from the latter by their steeper and shorter lingual protoconal slopes (Archibald, 1982; Luo, 1989, 1991). This is expressed by lower A/W-A ratios for *Protungulatum donnae* (Tables 40, 54) than for *Mimatuta morgoth* (Tables 53, 54) at McGuire Creek. Lower molars of *Mimatuta* are also similar in size to those of *Protungulatum donnae* (Tables 41, 42 and 55, 56), but are distinguished by the labial shift of the paraconids in the former (Archibald, 1982). This shift is accompanied by a deepening of the valley between the paraconid and metaconid (Luo, 1989). Also, lower molars of *M. morgoth* have on average slightly wider talonids than *P. donnae* (Archibald, 1982).

Upper and lower molars of *Protungulatum donnae* and *P. gorgun* from McGuire Creek are similar, but differ significantly in size (Tables 40, 41, 42 and 43, 44, 45). The size separation is great, with minimal overlap evident, and presumably denotes the presence of two separate species of *Protungulatum*. Lower molars of *P. donnae* also can be distinguished from those of *P. gorgun* by the development of a large anteriorly projecting para-conid in the latter (Van Valen, 1978).

A taxonomic problem is evident when comparing the McGuire Creek sample of *Protungulatum donnae* to the type and only specimen of *Protungulatum mckeeveri* (/P4-/M3, UCMP 121782). *P. mckeeveri* was distinguished from *P. donnae* by more reduced paraconids on /M1-3, longer /P4 and /M1-2 relative to width, and other minor differences in the proportions of the lower molars (Archibald, 1982). In contrast to the Bug Creek Anthills sample, the McGuire Creek sample referred to *P. donnae* has a range of variation in size that completely includes that of *P. mckeeveri* (Table 41). The McGuire Creek sample of *P. donnae* comes from a number of different localities in the upper Hell Creek Formation that probably differ slightly in age. Therefore, these two species may be present in the upper Hell Creek Formation at McGuire Creek, but are present at localities of slightly differing age. Alternatively, these species may have coexisted and were present at a single site. It is also possible that *P. donnae* and *P. mckeeveri* are synonymous, and data presented below suggest that this is probably true. A dentary fragment (UCMP 132595) here referred to *P. donnae* contains a /P4 of the size and proportions appropriate for *P. donnae* and an /M1 of the size and proportions of /M1's of *P. mckeeveri* (Tables 41, 42). The few isolated molars collected from the same site as UCMP 132595 are referable to *P. donnae*.

Another locality, V87098, produced a dentary fragment (UCMP 133838, Table 42) of *Protungulatum donnae* along with isolated molars that could be referred to *P. mckeeveri*. Other than slight differences in size and proportion, the only character that separates the two species is the slightly more reduced paraconids of *P. mckeeveri* (Archibald, 1982). A brief survey of the UCMP collections from Bug Creek Anthills, the type locality of *P. don-nae*, shows that a dentary fragment (UCMP 105006) has more reduced paraconids than *P. mckeeveri*. Given the data presented above, I suspect that *P. donnae* and *P. mckeeveri* are synonymous. However, a comparative study of *P. mckeeveri* and a large sample of "condylarths" from Bug Creek Anthills is needed to settle the matter.

*Protungulatum gorgun* Van Valen, 1978
Tables 43-45; Figures 24-26

*Protungulatum gorgun* Van Valen, 1978, p. 52.

*Type.* AMNH 35987, right mandible with /M2 (Van Valen, 1978, pl. 1, fig. 4).

*Type locality.* Harbicht Hill, Hell Creek Formation, Montana.

*Referred specimens.* Associated teeth /P4-/M3 (evidently from a dentary that was destroyed during screenwashing), UCMP 134622; dentary with /M3, UCMP 132318, from loc. V87028. One M1/, 1 M2/, 1 /M1 from loc. V87029. One /M2 from loc. V87034. One M1/, 2 M2/'s from loc. V87035. Dentary with /P2-3, /M1-2, UCMP 134558; 1 /M2 from loc. V87037. Dentary with /M2, UCMP 133064; 1 /M2, 1 M2/ from loc. V87038. One /M2, 1 M3/ from loc. V87049. Dentary with /M2, UCMP 133126, from loc. V87067. One M2/ from loc. V87071. One M2/ from loc. V87077. Dentary with /M1-2, UCMP 132806, from loc. V87084. One /M2 from loc. V87088. One M3/, 1 M2/ from loc. V87098. Dentary with /M2-3, UCMP 133817, from loc. V87107.

Table 43. Measurements of associated lower molars and premolars of *Protungulatum gorgun* from McGuire Creek localities

| Locality | Specimen | Tooth Site | Length | W/Tri | W/Tal |
|---|---|---|---|---|---|
| V87037 | 134558 | /P2 | 2.97 | 1.42 (width) | |
| | 134558 | /P3 | 3.24[a] | 1.76 (width) | |
| | 134558 | /M1 | 4.64 | 3.17 | 3.32 |
| | 134558 | /M2 | 4.96 | 3.90 | 3.65 |
| V87028 | 134622 | /P4 | 4.38 | 2.60 (width) | |
| | 134622 | /M1 | 4.43 | 3.08 | 3.06 |
| | 134622 | /M2 | 4.62 | 3.80 | 3.43 |
| | 134622 | /M3 | 5.82 | 3.60 | 2.88 |
| V87124 | 132436 | /M2 | 5.05 | 3.85 | 3.69 |
| | 132436 | /M3 | 5.57 | 3.57 | 2.92 |
| V87107 | 133817 | /M2 | — | 3.93 | 3.80 |
| | 133817 | /M3 | 6.16 | 3.50 | 2.83 |
| V87115 | 133837 | /M1 | 4.09 | 2.90 | 2.95 |
| | 133837 | /M2 | 5.00 | 3.80 | 3.50 |

a: Approximate measurement.

---

Dentary with /M1-2, UCMP 133837, from loc. V87115. One M2/ from loc. V87119. Dentary with /M2-3, UCMP 132436, from loc. V87124.

*Localities.* UCMP locs. V87028, V87029, V87034, V87035, V87037, V87038, V87049, V87067, V87071, V87077, V87084, V87088, V87098, V87107, V87115, V87119, and V87124.

*Distribution.* Upper Hell Creek Formation, Montana (Puercan).

*Description.* Of all the "condylarths" known to occur in the upper Hell Creek Formation in eastern Montana, *Protungulatum gorgun* is undoubtedly the poorest known. Van Valen's (1978) diagnosis of *P. gorgun* is brief, and he provides little additional morphological description. The only other description of *P. gorgun* was provided by Lupton et al. (1980), who described two dentary fragments collected from Chris' Bonebed, a Milwaukee Public Museum locality in the upper Hell Creek Formation in McCone County. The upper dentition of *P. gorgun* has never been described.

Fossils from McGuire Creek are referred to *Protungulatum gorgun* on the basis of comparisons to an epoxy cast of the type (AMNH 35987), an isolated /M2 from Harbicht Hill. A number of dentaries from McGuire Creek referable to *P. gorgun* contain the /M2 with other parts of the lower dentition (Table 43). These fossils form the basis for the description of the lower dentition presented below.

The portion of the mandible anterior to /P4 of *Protungulatum gorgun* is preserved only in UCMP 134558 (Figure 24). UCMP 134558 has two mental foramina, one below the /P1-/P2 diastema, and a second below the posterior root of /P3. Only the root of

Table 44. Measurements of associated and isolated lower molars of *Protungulatum gorgun* from McGuire Creek localities, and the type (AMNH 35987) from Harbicht Hill (V71203)

| Tooth Site | Number | Observed Range | Mean |
|---|---|---|---|
| /M1 | | | |
| Length | 4 | 4.09-4.64 | 4.36 |
| Width-Trigonid | 4 | 2.90-3.25 | 3.10 |
| Width-Talonid | 4 | 2.95-3.40 | 3.22 |
| W-Tri/W-Tal | 4 | .95-1.01 | .98 |
| Length/W-Tal | 4 | 1.26-1.45 | 1.37 |
| /M2 | | | |
| Length | 12 | 4.62-5.38 (5.28[a]) | 5.08 |
| Width-Trigonid | 11 | 3.69-4.17 (4.01[a]) | 3.91 |
| Width-Talonid | 11 | 3.43-3.94 (3.70[a]) | 3.65 |
| W-Tri/W-Tal | 11 | 1.03-1.11 (1.08[a]) | 1.08 |
| Length/W-Tal | 11 | 1.34-1.47 (1.43[a]) | 1.39 |
| /M3 | | | |
| Length | 4 | 5.34-6.16 | 5.72 |
| Width-Trigonid | 4 | 3.50-3.63 | 3.58 |
| Width-Talonid | 4 | 2.83-3.04 | 2.92 |
| W-Tri/W-Tal | 4 | 1.19-1.25 | 1.23 |
| Length/W-Tal | 4 | 1.76-2.18 | 1.97 |

(a): Measurements of type specimen from cast of AMNH 35987, an isolated /M2. Measurements of type not included in means.

---

the canine is preserved, and it is larger than the roots of /P1 and /P2. The canine root is recurved under the root of /P1. /P1 is missing, but its alveolus is angled posteriorly, indicating that /P1 was slightly procumbent. There is a short diastema between /P1-/P2 and one nearly twice as long between /P2-/P3. The remaining premolars have two roots. /P2 has a large central cusp with weakly developed anterior and posterior midline ridges. Also present is a small posterior basal cusp, but an anterior basal cusp is lacking. /P3 has a large central cusp that is higher and more robust than that of /P2. A small anterior basal cusp is present and is shifted lingual to the midline of the tooth. The posterior portion of the /P3 is missing because of breakage.

Only one dentary (UCMP 134622) has a /P4 (Figure 25D-F), and it is moderately worn. The /P4 has a tall, robust protoconid with a small metaconid placed low on its side. The paraconid is large, slightly procumbent, well separated from the protoconid, and positioned on the anterior basal part of the tooth lingual to the midline. /P4 is slightly shorter than /M1 on UCMP 134622 (Table 43), and the widest part of the tooth is at the junction of trigonid and talonid. The talonid is short but wide, with a low and small cusp situated at its posterior edge in the position normally occupied by a hypoconid.

Table 45. Measurements of isolated upper molars of *Protungulatum gorgun* from McGuire Creek localities

| Locality | Specimen | Tooth Site | Length | W-A | W-P |
|---|---|---|---|---|---|
| V87035 | 132505 | M1/ | — | — | 6.03 |
| V87029 | 132345 | M1/ | — | — | 6.30 |
| V87035 | 132507 | M2/ | 5.66 | 6.45 | 7.06 |
| V87035 | 132502 | M2/ | 5.09 | 6.07[a] | 6.52 |
| V87077 | 133525 | M2/ | 5.00[a] | 6.53[a] | 7.08 |
| V87119 | 132439 | M2/ | — | 6.52 | 6.93 |
| V87071 | 133247 | M2/ | — | — | 6.91 |
| V87038 | 132117 | M2/ | 4.78 | 6.20 | 6.49 |
| V87049 | 133145 | M3/ | 4.70 | 5.73 | 5.29 |

a: Approximate measurement.

In comparison to other "condylarths" found in the upper Hell Creek Formation, the premolars of *Protungulatum gorgun* are shorter and narrower than those of *Ragnarok*, but are longer and wider than those of *Protungulatum donnae* and *Mimatuta morgoth*. In basic morphology, the premolars of *Protungulatum gorgun* and *P. donnae* are similar, but the /P2-/P3 diastema is relatively longer in *P. gorgun*. Also, the /P4 of *P. donnae* is widest through the metaconid-protoconid (Archibald, 1982), while the /P4 of *P. gorgun* is widest more posteriorly at the talonid-trigonid contact. The /P4 talonid of *P. gorgun* is better developed (relatively wider and longer) and has a relatively larger hypoconid.

The lower molars of *Protungulatum gorgun* and *P. donnae* are similar morphologically, with a few exceptions. Lower molars of *P. gorgun* are larger, but similar in proportions (Tables 41, 44). The trigonid cusps of *P. gorgun* are relatively more bulbous and the paraconid and metaconid are more separated (Figures 24, 26). As in *P. donnae*, the paraconid of *P. gorgun* projects anteriorly, but does so to a greater degree. Unlike *P. donnae*, the paraconid in *P. gorgun* is usually shifted labially in relation to a line extended through the apices of the entoconid and metaconid. The hypoconid is larger relative to the other talonid cusps in *P. gorgun* (Figures 24, 26).

In comparison to *Ragnarok*, lower molars of *Protungulatum gorgun* are smaller and their trigonids are higher relative to the talonid. The trigonid and talonid cusps are less bulbous and the trigonid walls are much less expanded in *P. gorgun*. Also, the paraconid projects anteriorly in *P. gorgun* and does not (or projects very slightly) in *Ragnarok*.

The upper dentition of *Protungulatum gorgun* has never been described. Unfortunately, upper molars were not found in association with any of the mandibles referred to *P. gorgun* from McGuire Creek. Many isolated molars from McGuire Creek are similar to *P. donnae*, but are significantly larger and in the expected size range of *P. gorgun*. Many of these upper molars were collected from sites producing portions of

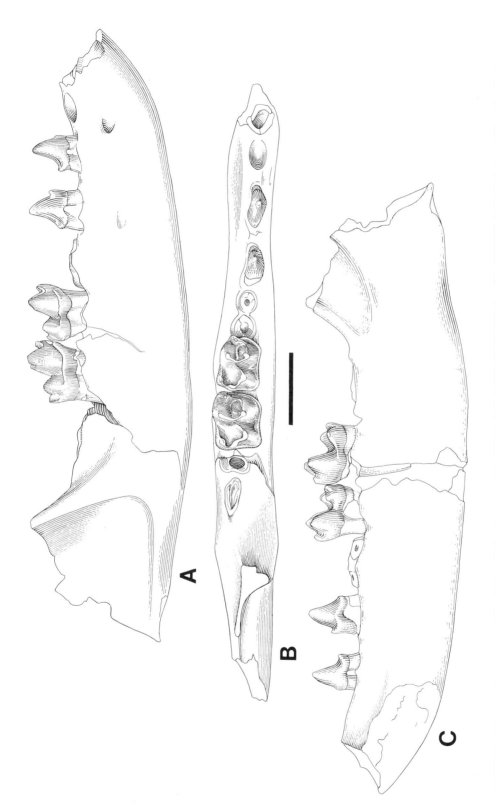

FIGURE 24. *Protungulatum gorgun* Van Valen. Right dentary with /P2-3, /M1-2, UCMP 134558, locality V87037: (A) labial view, (B) occlusal view, (C) lingual view. Scale bar = 5 mm.

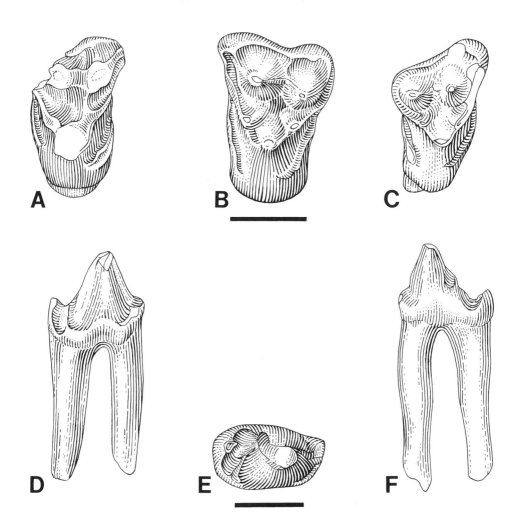

FIGURE 25. *Protungulatum gorgun* Van Valen. (A) Isolated left M1/, UCMP 132345, locality V87029, occlusal view. (B) Isolated left M2/, UCMP 132507, locality V87035, occlusal view. (C) Isolated right M3/, UCMP 133145, locality V87049, occlusal view. Scale bar = 4 mm. Isolated right /P4 (associated with /M1-3 but molars not shown), UCMP 134622, locality V87028: (D) labial view, (E) occlusal view, (F) lingual view. Scale bar = 3 mm.

lower dentitions that can be referred to *P. gorgun* with confidence. The only feature of the upper dentition mentioned in the diagnosis of *P. gorgun* concerns the protocone, "upper molars with rather massive protocone" (Van Valen, 1978, p. 52). The McGuire Creek fossils have fairly large protocones, but this character exhibits a wide range of variation. Therefore, reference of upper molars from McGuire Creek to *P. gorgun* must be regarded as tentative.

Both M1/'s from McGuire Creek referred to *Protungulatum gorgun* lack their parastylar region (Figure 25A). M1/'s of *P. gorgun* have a relatively larger protocone and a weaker postcingulum and hypocone than *P. donnae*. In other respects, they are basically a larger

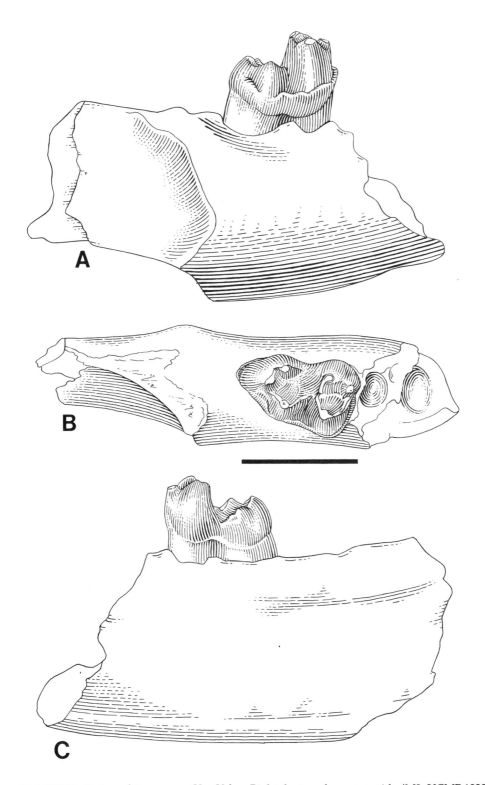

FIGURE 26. *Protungulatum gorgun* Van Valen. Right dentary fragment with /M3, UCMP 132318, locality V87028: (A) labial view, (B) occlusal view, (C) lingual view. Scale bar = 5 mm.

version of M1/'s of *P. donnae*. The M2/ protocone of *P. gorgun* is high and large, which causes the labial and lingual protoconal slope to be steeper and longer than in M2/'s of *P. donnae* (Figure 25B). Also, the M2/ of *P. gorgun* is relatively wider. M3/'s of *P. gorgun* are larger than those of *P. donnae*, but otherwise are similar (Figure 25C). Dental measurements of *P. gorgun* are presented in Table 45. Upper premolars of *P. gorgun* were not identified.

*Oxyprimus* Van Valen, 1978<br>*Oxyprimus erikseni* Van Valen, 1978<br>Tables 46-48; Figure 27

*Oxyprimus erikseni* Van Valen, 1978, p. 53.

   *Type.* UMVP 1561, isolated /M1 (Van Valen, 1978, pl. 1, fig. 5).

   *Type locality.* Harbicht Hill, Hell Creek Formation, Montana.

   *Referred specimens.* Dentary with /M3, UCMP 132453; 1 M2/ from loc. V86031. One M1/ from loc. V87028. Dentary with /M2-3, UCMP 132348; 1 M2/ from loc. V87029. One M1/ from loc. V87034. One /Mx from loc. V87035. One M1/ from loc. V87036. Dentary with /P2, /P4-/M2, UCMP 133068; 1 M1/ from loc. V87038. One /P4, 1 P4/ from loc. V87049. One M2/, 1 /M3 from loc. V87051. Dentary with /M1-3, UCMP 132350; 2 Mx/'s from loc. V87052. One M1/, 1 /M1, 2 /M2, 1 /M3, 2 M3/'s from loc. V87071. Two P4/'s, 1 DP4/, 2 M1/'s, 1 M2/, 1 M3/, 2 /M1's, 1 /M2, 1 /M3, 2 /Mx's from loc. V87072. One M1/, 1 M3/, 1 /M1 from loc. V87074. One /M3 from loc. V87084. One /M2, 1 /M3 from loc. V87095. One P4/, 2 M1/'s, 2 /P4's, 1 /M3, 1 /Mx from loc. V87098. Three M1/'s, 2 /M3's from loc. V87151. One M1/, 2 Mx/'s, 1 /M3 from loc. V87153.

   *Localities.* UCMP locs. V86031, V87028, V87029, V87034, V87035, V87036, V87038, V87049, V87051, V87052, V87071, V87072, V87074, V87084, V87095, V87098, V87151, and V87153.

   *Distribution.* Upper Hell Creek and lower Tullock formations, Montana (both Puercan); possibly Ravenscrag Formation, Saskatchewan (Puercan).

   *Description.* Reference of lower molars and premolars from McGuire Creek to

Table 46. Measurements of associated lower molars and premolars of *Oxyprimus erikseni* from McGuire Creek localities

| Locality | Specimen | Tooth Site | Length | W-Tri | W-Tal |
|---|---|---|---|---|---|
| V87038 | 133068 | /P2 | 2.22 | 1.02 (width) | |
| | 133068 | /P4 | 3.00 | 1.62 (width) | |
| | 133068 | /M1 | 3.03 | 2.04 | 2.16 |
| | 133068 | /M2 | 3.35 | 2.45 | 2.17 |
| V87052 | 132350 | /M1 | 3.18 | 2.24 | 2.28 |
| | 132350 | /M2 | 3.40 | 2.68 | 2.55 |
| | 132350 | /M3 | 4.13 | 2.58 | 2.02 |
| V87029 | 132348 | /M2 | 3.64 | 2.68 | 2.73 |
| | 132348 | /M3 | 4.14 | 2.44 | 2.00 |

*Oxyprimus erikseni* is based on comparison of /M1's to a cast of the type, UMVP 1561, an isolated /M1 from Harbicht Hill. Fortunately, two dentary fragments from McGuire Creek preserved /M1's with other parts of the dental arcade (Table 46; Figure 27C), and these specimens provide the basis for identification of isolated /P4's and /M2-3's. Dental measurements are given in Tables 46 and 47.

Upper and lower molars of *Oxyprimus erikseni* have not been found in association, and upper molars and premolars from Harbicht Hill, the type locality, have never been described. Isolated upper molars and premolars from McGuire Creek are referred to *O. erikseni* on the basis of comparison of these fossils with casts of *O. galadrielae*, a very similar species from Mantua Lentil. The upper dentitions of *O. galadrielae* and *O. erikseni* are difficult to distinguish (see discussion below), but upper molars and premolars referred to *O. erikseni* were collected from sites that produced portions of lower dentitions referred confidently to *O. erikseni*. The lower dentitions of *O. galadrielae* and *O. erikseni* are similar, but can be separated by a few characters (see discussion below). Therefore, although reference of these specimens from McGuire Creek to *O. erikseni* is probably correct, uncertainty exists. Dental measurements are given in Table 48.

The dentition of *Oxyprimus erikseni* was described by Archibald (1982), and his comments apply equally well to fossils referred to *O. erikseni* from McGuire Creek. A few modifications and additions are necessary. Archibald's (1982) identification of the tooth sites of three of the upper molars from Worm Coulee 1 (V74111) probably is incorrect. This conclusion is based on my comparison of these molars to specimens of *Oxyprimus* from McGuire Creek. UCMP 116513 and 116515 are actually M1/'s not M2/'s and UCMP 116511, as suspected by Archibald (1982), is a DP4/, not an M1/. Therefore, the M1/ and M2/ of *O. erikseni* require additional description. M1/'s and M2/'s of *O. erikseni* are similar to those of *Protungulatum donnae* (Archibald, 1982), but smaller. In contrast to *P. donnae*, *Oxyprimus erikseni* has a smaller parastylar lobe, a relatively larger hypocone, and the lingual slope of the protocone is steeper (Figure 27A-B). Also, weak internal conular wings are more commonly developed in *O. erikseni* than in *Protungulatum donnae*.

*Discussion.* A potential taxonomic problem exists in separating similar-sized species of *Oxyprimus* from Harbicht Hill and Mantua Lentil. The dentitions of *O. erikseni* from Harbicht Hill, Hell Creek Formation, Montana, and *O. galadrielae* and *O. putorius* from Mantua Lentil, Polecat Bench Formation, Wyoming, are similar.

Van Valen's (1978) diagnosis of each species is so brief that it became necessary to evaluate their validity. Identification of species is important, because regional biostratigraphic and biochronologic correlations are based in part on species-level differences within *Oxyprimus*. Van Valen (1978) only noted differences in metaconid shape as a means of separating *O. galadrielae* and *O. erikseni*. According to the diagnosis of each, *O. erikseni* has a conical molar metaconid and *O. galadrielae* has an elongate molar metaconid. Close scrutiny of the two species reveals that this is not the case. Molar metaconids of *O. erikseni* were found to be proportionately longer than or equal in diameter to those of *O. galadrielae*. Measurements were taken from the apex of the metaconid of /M1 to the transverse groove separating this cusp from the paraconid, and then were

Table 47. Measurements of associated and isolated lower molars and premolars of *Oxyprimus erikseni* from Harbicht Hill and McGuire Creek localities and *O. galadrielae* from Mantua Lentil[a]

| | McGuire Creek O. erikseni | Harbicht Hill O. erikseni | Mantua Lentil O. galadrielae |
|---|---|---|---|
| **/P4 Length** | | | |
| Number | 3 | | 3 |
| Observed Range | 3.00-3.45 | | 2.80-2.99 |
| Mean | 3.17 | | 2.90 |
| **/P4 Width** | | | |
| Number | 3 | | 3 |
| Observed Range | 1.62-1.80 | | 1.68-1.86 |
| Mean | 1.73 | | 1.77 |
| **/M1 Length** | | | |
| Number | 5 | 2 | 4 |
| Observed Range | 3.03-3.38 | 3.35-3.36 | 3.21-3.26 |
| Mean | 3.23 | 3.36 | 3.23 |
| **/M1 W-Tri** | | | |
| Number | 5 | 2 | 4 |
| Observed Range | 1.97-2.24 | 2.14-2.22 | 2.09-2.21 |
| Mean | 2.07 | 2.18 | 2.14 |
| **/M1 W-Tal** | | | |
| Number | 5 | 2 | 4 |
| Observed Range | 1.91-2.28 | 2.27-2.28 | 2.27-2.45 |
| Mean | 2.13 | 2.28 | 2.38 |
| **/M1 W-Tri/W-Tal** | | | |
| Number | 5 | 2 | 4 |
| Observed Range | .94-1.03 | .94- .98 | .86 - .92 |
| Mean | .97 | .96 | .90 |
| **/M1 Length/W-Tal** | | | |
| Number | 5 | 2 | 4 |
| Observed Range | 1.39-1.76 | 1.47-1.48 | 1.31-1.43 |
| Mean | 1.52 | 1.48 | 1.36 |
| **/M2 Length** | | | |
| Number | 7 | | 3 |
| Observed Range | 3.35-3.64 | | 3.43-3.74 |
| Mean | 3.53 | | 3.60 |
| **/M2 W-Tri** | | | |
| Number | 7 | | 3 |
| Observed Range | 2.40-2.82 | | 2.64-2.73 |
| Mean | 2.59 | | 2.67 |
| **/M2 W-Tal** | | | |
| Number | 7 | | 3 |
| Observed Range | 2.18-2.73 | | 2.56-2.71 |
| Mean | 2.43 | | 2.64 |

| | McGuire Creek *O. erikseni* | Harbicht Hill *O. erikseni* | Mantua Lentil *O. galadrielae* |
|---|---|---|---|
| **/M2 W-Tri/W-Tal** | | | |
| Number | 7 | | 3 |
| Observed Range | .98-1.13 | | 1.00-1.03 |
| Mean | 1.07 | | 1.01 |
| **/M2 Length/W-Tal** | | | |
| Number | 7 | | 3 |
| Observed Range | 1.33-1.65 | | 1.29-1.41 |
| Mean | 1.46 | | 1.36 |
| **/M3 Length** | | | |
| Number | 7 | | 3 |
| Observed Range | 3.83-4.52 | | 3.76-4.31 |
| Mean | 4.17 | | 3.96 |
| **/M3 W-Tri** | | | |
| Number | 7 | | 3 |
| Observed Range | 2.28-2.80 | | 2.37-2.53 |
| Mean | 2.50 | | 2.46 |
| **/M3 W-Tal** | | | |
| Number | 7 | | 3 |
| Observed Range | 1.76-2.08 | | 1.96-2.10 |
| Mean | 1.98 | | 2.03 |
| **/M3 W-Tri/W-Tal** | | | |
| Number | 7 | | 3 |
| Observed Range | 1.15-1.35 | | 1.20-1.22 |
| Mean | 1.27 | | 1.21 |
| **/M3 Length/W-Tal** | | | |
| Number | 7 | | 3 |
| Observed Range | 1.89-2.18 | | 1.84-2.05 |
| Mean | 2.11 | | 1.95 |

a. Underlined measurements from Harbicht Hill were taken from an epoxy cast of the type of *Oxyprimus erikseni*, UMVP 1561, an isolated /M1.

compared with trigonid length. This ratio measures the elongation of the metaconid relative to length of the trigonid. *Oxyprimus erikseni* (n=3, including an epoxy cast of the type) has metaconid lengths of .78-.83 and elongation ratios of .44-.48. For *O. galadrielae* (from casts of PU 16712, PU 16863) these values are .71-.73 and .44-.45, respectively. On the basis of this small sample, the /M1 metaconid of *O. erikseni* appears to be equal to or slightly more elongate than metaconids of *O. galadrielae*. Inspection of /M2's and /M3's referred to *O. erikseni* from McGuire Creek and *O. galadrielae* from Mantua Lentil show the same trend. Therefore, molar metaconid elongation has questionable taxonomic value, and the diagnoses of *O. galadrielae* and *O. erikseni* are changed below to reflect this.

Because of the inadequacy of diagnoses of *Oxyprimus galadrielae* and *O. erikseni*, the possibility that these species may be synonymous was evaluated. Comparison of casts of

Table 48.  Measurements of isolated upper molars of *Oxyprimus erikseni* from various McGuire Creek localities compared with measurements of casts of the type specimens of *O. galadrielae* and *O. putorius* from Mantua Lentil

| Tooth Site M1/ | Number | Observed Range | Mean |
|---|---|---|---|
| Length | 7 | 3.17-3.93 (3.36[a])(3.11[b]) | 3.42 |
| Width-Anterior | 8 | 4.23-4.64 (4.45[a])(3.62[b]) | 4.42 |
| Width-Posterior | 8 | 4.32-4.74 (4.45[a])(3.96[b]) | 4.53 |
| Length/Width-P | 7 | .72-.83  (.76[a]) (.79[b]) | .75 |
| M2/ | | | |
| Length | 4 | 3.64-3.90 (3.59[a]) | 3.78 |
| Width-Anterior | 4 | 5.08-5.24 (4.98[a])(4.31[b])[c] | 5.17 |
| Width-Posterior | 4 | 5.36-5.68 (5.23[a])(4.73[b])[c] | 5.49 |
| Length/Width-P | 4 | .68-.72  (.69[a]) | .69 |
| M3/ | | | |
| Length | 3 | 3.07-3.23 (2.51[a]) | 3.15 |
| Width-Anterior | 3 | 3.56-3.95 (3.99[a]) | 3.69 |
| Width-Posterior | 3 | 3.27-3.51 (3.28[a]) | 3.39 |
| Length/Width-P | 3 | .91-.96  (.77[a]) | .93 |

a. *Oxyprimus galadrielae*, PU 16866.
b. *Oxyprimus putorius*, PU 16704.
c. Approximate measurement.

---

the lower dentition of *O. galadrielae* with specimens referred to *O. erikseni* from McGuire Creek and Harbicht Hill (including a cast of the type) revealed a few differences. *O. erikseni* has slightly narrower molar talonids than does *O. galadrielae*, both in proportion and in absolute size (Table 47). Also, in *O. erikseni* /P4's and /M1's are subequal in length, whereas in *O. galadrielae* /M1's are significantly longer than /P4's (Table 47). In most cases, *O. erikseni* has a larger /P4 metaconid than *O. galadrielae* although there is some degree of overlap. Otherwise the lower dentitions of these two species are similar.

Upper molars of *Oxyprimus erikseni* and *O. galadrielae* are also similar. Comparison of the type of *O. galadrielae* (PU 16866) to upper molars referred to *O. erikseni* from McGuire Creek reveal no significant differences. M2/'s from McGuire Creek appear to be slightly more transverse, but the difference is small, and proportionately they are identical (Table 48).

A second species from Mantua Lentil, *Oxyprimus putorius*, described briefly by Van Valen (1978), may be valid. Upper molars of *O. putorius*, known only from the type specimen (PU 16704), are less transverse than those of both *O. galadrielae* and *O. erikseni*

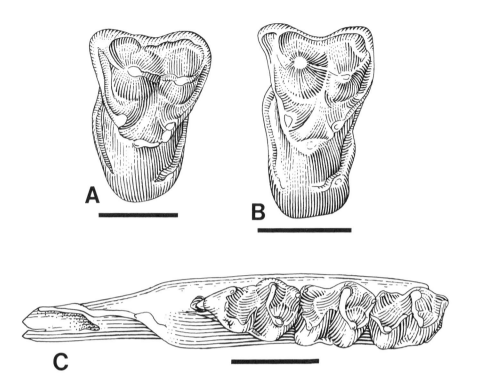

FIGURE 27. *Oxyprimus erikseni* Van Valen. (A) Isolated left M1/, UCMP 133447, locality V87072, occlusal view. Scale bar = 2.5 mm. (B) Isolated left M2/, UCMP 134698, locality V87072, occlusal view. Scale bar = 3 mm. (C) Right dentary fragment with /M1-3, UCMP 132350, locality V87052, occlusal view. Scale bar = 3 mm.

(Van Valen, 1978), although proportionately there is little difference among upper molars of the three species (Table 48). The only known specimen of a lower dentition referred to *O. putorius* (PU 21099) is so poorly preserved that it cannot be used for comparative purposes. More material is needed to assess the validity of *O. putorius*.

In summary, *O. galadrielae* and *O. erikseni* are similar, but appear to be valid species that can be distinguished by differences in their lower dentitions. *O. putorius* is poorly known, and the validly of this species is questionable. A revised diagnosis of each species is given below:

*Oxyprimus erikseni*: /P4 metaconid usually large, narrower molar talonids than *O. galadrielae*, /P4 and /M1 subequal in length, upper molars more transverse than *O. putorius*.

*Oxyprimus galadrielae*: /P4 metaconid small, wider molar talonids than *O. erikseni*, /M1 longer than /P4, upper molars more transverse than *O. putorius*.

*Oxyprimus putorius*: upper molars less transverse than *O. erikseni* and *O. galadrielae*.

Subfamily LOXOLOPHINAE Van Valen, 1978
*Ragnarok* Van Valen, 1978
*Ragnarok nordicum* (Jepsen, 1930) Van Valen, 1978
Tables 49-52; Figures 28-30

*Loxolophus nordicus* Jepsen, 1930, p. 501-502.
*Ragnarok harbichti* Van Valen, 1978, p. 56.

*Type.* PU 13285, right maxilla with M1-3/ (Jepsen, 1930, pl. IV, fig. 1).

*Type locality.* Mantua Lentil, Polecat Bench Formation (Fort Union Formation), Wyoming.

*Referred specimens.* Dentary with /M2 talonid, /M3, UCMP 132594, from loc. V84151. Dentary with /M2-3, UCMP 132444; dentary with /P4-/M3, UCMP 132458; 1 M1/, 1 /M1, 2 /M2's, 2 /M3's from loc. V86031. Dentary with /M2-3, UCMP 132435; 1 /P4 from loc. V87033. Dentary with /M3, UCMP 132473; 1 /P4 from loc. V87034. Dentary with /M3, UCMP 132500, from loc. V87035. Two /M2's, 2 /M3's, 1 /Mx from loc. V87037. Associated /M1 and /M2 from loc. V87050. Dentary with /C, /P1-2, /P4, /M1-2, /M3 trigonid, UCMP 134797; maxilla fragment with P1/-M3/, UCMP 134694; maxilla fragment with M1-3/, UCMP 134693; 1 P4/ from loc. V87072. Dentary with /C, alveoli for /P1-3, UCMP 132804, from loc. V87088. Dentary with /M2 talonid, /M3, UCMP 132426, from loc. V87109. Dentary with /M2-3, UCMP 134592; dentary with /M2 talonid, /M3, UCMP 134591, from loc. V88041. One to three isolated teeth from 16 other localities.

*Localities.* UCMP locs. V84151, V86031, V87029, V87030, V87033, V87034, V87035, V87036, V87037, V87038, V87049, V87050, V87072, V87073, V87074, V87077, V87078, V87082, V87088, V87098, V87109, V87117, V87124, V87152, V87153, V88041, and V88044.

*Distribution.* Upper Hell Creek Formation, Montana, and Polecat Bench Formation (Fort Union Formation), Wyoming (both Puercan).

*Discussion and Description.* When Van Valen (1978) erected the genus *Ragnarok*, it included the genotypic species *R. harbichti* from Harbicht Hill, upper Hell Creek Formation, Montana; *R. nordicum* (*Loxolophus nordicus*, Jepsen, 1930) from Mantua Lentil, Polecat Bench Formation, Wyoming; and *R. wovokae* from Rock Bench, Wyoming. Later, Archibald (1982) named a new species, *R. engdahli*, based on isolated teeth from Worm Coulee 1 (V74111), lower Tullock Formation, Montana. Analysis of the type specimens and the hypodigm of these four species indicates that fossils from McGuire Creek probably represent a single population of *Ragnarok* that is similar to *R. nordicum* and *R. harbichti*. According to their diagnoses, *Ragnarok nordicum* and *R. harbichti* are similar and differ only in size of the molar talonids (see Van Valen, 1978). Specimens of *R. harbichti* from Harbicht Hill have slightly smaller /M1-2 talonids on average than the hypodigm of *R. nordicum* from Mantua Lentil, but proportionally (L/W-Tal) only the /M2 talonid of *R. nordicum* appears larger (Table 49). Because talonid width and total length of the /M2 appear to exhibit the greatest difference between the samples from the type localities of *R. harbichti* and *R. nordicum*, these values were subjected to T-tests. The differences in means of /M2 talonid width and /M2

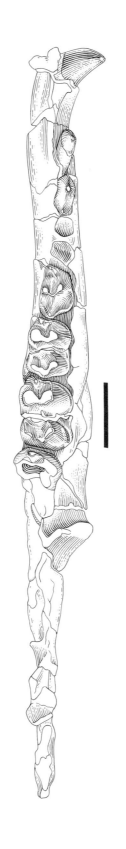

FIGURE 28. *Ragnarok nordicum* Jepsen. Right dentary with /C, /P1-2, /P4, /M1-2, partial trigonid of /M3, UCMP 134797, locality V87072, occlusal view. Scale bar = 5 mm.

Table 49. Measurements of associated and isolated lower molars and premolars of *Ragnarok nordicum* (formerly *R. harbichti*) from Harbicht Hill (V71203) and *R. nordicum* from McGuire Creek localities and Mantua Lentil[a]

| | McGuire Creek | Harbicht Hill | Mantua Lentil |
|---|---|---|---|
| /P4 Length | | | |
| Number | 6 | 1 | 1 |
| Observed Range | 4.62-4.93 | 4.22 | 4.60 |
| Mean | 4.74 | 4.22 | 4.60 |
| /P4 Width | | | |
| Number | 6 | 1 | 1 |
| Observed Range | 2.82-3.30 | 3.20 | 2.81 |
| Mean | 3.07 | 3.20 | 2.81 |
| /M1 Length | | | |
| Number | 7 | 1 | 3 |
| Observed Range | 4.78-5.13 | 4.94 | 5.13-5.26 |
| Mean | 4.97 | 4.94 | 5.18 |
| /M1 W-Tri | | | |
| Number | 7 | 1 | 3 |
| Observed Range | 3.62-3.94 | 3.94 | 3.60-4.08 |
| Mean | 3.83 | 3.94 | 3.86 |
| /M1 W-Tal | | | |
| Number | 7 | 2 | 3 |
| Observed Range | 3.71-4.06 | 3.65-4.00 | 3.80-4.35 |
| Mean | 3.85 | 3.83 | 4.06 |
| /M1 W-Tri/W-Tal | | | |
| Number | 7 | 1 | 3 |
| Observed Range | .97-1.02 | .99 | .89-1.03 |
| Mean | 1.00 | .99 | .95 |
| /M1 Length/W-Tal | | | |
| Number | 7 | 1 | 3 |
| Observed Range | 1.26-1.33 | 1.24 | 1.21-1.35 |
| Mean | 1.29 | 1.24 | 1.28 |
| /M2 Length | | | |
| Number | 12 | 4 | 4 |
| Observed Range | 5.29-5.89 | 5.80-5.92 | 5.26-5.76 |
| Mean | 5.66 | 5.84 | 5.50 |
| /M2 W-Tri | | | |
| Number | 12 | 4 | 4 |
| Observed Range | 4.58-5.38 | 5.11-5.20 | 4.66-5.42 |
| Mean | 5.00 | 5.16 | 5.09 |
| /M2 W-Tal | | | |
| Number | 12 | 4 | 4 |
| Observed Range | 4.07-4.92 | 4.46-4.82 | 4.63-5.43 |
| Mean | 4.34 | 4.61 | 4.96 |

|  | McGuire Creek | Harbicht Hill | Mantua Lentil |
|---|---|---|---|
| /M2 W-Tri/W-Tal | | | |
| Number | 12 | 4 | 4 |
| Observed Range | 1.06-1.19 | 1.08-1.14 | 1.00-1.09 |
| Mean | 1.13 | 1.12 | 1.03 |
| /M2 Length/W-Tal | | | |
| Number | 12 | 4 | 4 |
| Observed Range | 1.17-1.33 | 1.20-1.30 | 1.06-1.15 |
| Mean | 1.28 | 1.27 | 1.11 |
| /M3 Length | | | |
| Number | 12 | 2 | 4 |
| Observed Range | 5.98-6.98 | 6.27-6.32 | 6.24-7.02 |
| Mean | 6.40 | 6.30 | 6.52 |
| /M3 W-Tri | | | |
| Number | 12 | 2 | 4 |
| Observed Range | 4.11-4.77 | 4.20-4.43 | 4.14-4.77 |
| Mean | 4.45 | 4.32 | 4.55 |
| /M3 W-Tal | | | |
| Number | 12 | 2 | 4 |
| Observed Range | 3.20-3.82 | 3.40-3.59 | 3.28-3.76 |
| Mean | 3.52 | 3.50 | 3.61 |
| /M3 W-Tri/W-Tal | | | |
| Number | 12 | 2 | 4 |
| Observed Range | 1.17-1.33 | 1.23-1.23 | 1.24-1.27 |
| Mean | 1.27 | 1.23 | 1.26 |
| /M3 Length/W-Tal | | | |
| Number | 12 | 2 | 4 |
| Observed Range | 1.68-1.94 | 1.76-1.84 | 1.67-1.93 |
| Mean | 1.82 | 1.80 | 1.81 |

a. Underlined measurements from Harbicht Hill were taken from a cast of the type specimen of *Ragnarok nordicum* (formerly *R. harbichti*) (AMNH 35983). Other measurements of *R. nordicum* from Harbicht Hill made from UCMP 136094 (/P4-/M2) and two casts (UMVP 2076, /M2-3; UMVP 1555, /M2). Measurements of *R. nordicum* from Mantua Lentil made from casts (PU 14475, /M1-3; PU 21123, /M2; PU 14488, /M3; PU 14473, /M3; PU 14174, /M1-2; PU 16720, /P4-/M3).

Table 50.  Measurements of associated lower molars and premolars of *Ragnarok nordicum* from McGuire Creek localities

| Locality | Specimen | Tooth Site | Length | W-Tri | W-Tal |
|---|---|---|---|---|---|
| V87072 | 134797 | /C | 3.53[a] | 2.55 (width) | |
| | 134797 | /P1 | 2.17 | 1.39 (width) | |
| | 134797 | /P2 | 3.40 | 1.93 (width) | |
| | 134797 | /P4 | 4.66 | 3.30 (width) | |
| | 134797 | /M1 | 5.13 | 3.94 | 4.06 |
| | 134797 | /M2 | 5.69 | 5.20 | 4.92 |
| | 134797 | /M3 | — | 4.60 | — |
| V86031 | 132458 | /P4 | 4.76 | 3.16 (width) | |
| | 132458 | /M1 | 5.01 | 3.86 | 3.78 |
| | 132458 | /M2 | 5.89 | 5.38[a] | 4.69 |
| | 132458 | /M3 | 6.95 | 4.74 | 3.82 |
| V86031 | 132444 | /M2 | 5.29 | 4.80 | 4.24 |
| | 132444 | /M3 | 6.50 | 4.57 | 3.82 |
| V87033 | 132435 | /M2 | 5.56 | 5.09 | 4.47 |
| | 132435 | /M3 | 6.38 | 4.67 | 3.63 |
| V88044 | 134591 | /M2 | — | — | 4.55 |
| | 134591 | /M3 | 6.21 | 4.50 | 3.70 |
| V88044 | 134592 | /M2 | 5.84 | 5.17 | 4.72 |
| | 134592 | /M3 | 6.07 | 4.41 | 3.46 |
| V87050 | 132307 | /M1 | 5.02 | 3.80 | 3.88 |
| | 132306 | /M2 | 5.78 | 4.94 | 4.34 |

a: Approximate measurement.

length between the two samples were not significant (/M2 length: t=1.93, df=6, p=.10; /M2 talonid width: t=1.88, df=6, .20>p>.10). The Mantua Lentil and Harbicht Hill samples cannot be confidently separated by size of the molar talonids, and the diagnosis of each species is inadequate. A comparison of the remainder of the upper and lower dentitions of *R. nordicum* and *R. harbichti* reveals a lack of morphological characters that separate the species. On the basis of available information, *R. nordicum* and *R. harbichti* are considered synonymous, with *R. nordicum* having priority. Therefore, the sample of *Ragnarok* from McGuire Creek is referred to *R. nordicum*. It should be noted, however, that the sample sizes available for study from Harbicht Hill and Mantua Lentil are small (n=4 or less; Table 49).

Middleton (1983) has proposed that *Ragnarok* is synonymous with *Baioconodon*, and his views have gained acceptance (Sloan et al., 1986; Smit et al., 1987; Archibald and Lofgren, 1990). But until this synonymy is formally published, I will continue to recognize *Ragnarok*.

Table 51. Measurements of associated upper molars and premolars of *Ragnarok nordicum* from locality V87072

| Specimen | Tooth Site | Length | W-A | W-P |
|----------|-----------|--------|-----|-----|
| 134694 | P2/ | 3.22 | 2.05 (width) | |
| 134694 | P3/ | 3.80 | — | — |
| 134694 | P4/ | 4.30 | 5.67 | 5.89 |
| 134694 | M1/ | 5.50 | 6.73 | 6.89 |
| 134694 | M2/ | 5.92 | 8.13 | 8.26 |
| 134694 | M3/ | 5.42 | 6.84 | 6.15 |
| 134693 | M1/ | 5.01 | 6.56 | 6.91 |
| 134693 | M2/ | 5.84 | 8.04 | 8.47 |
| 134693 | M3/ | 5.94 | 6.28 | 5.57 |

The dentition of *Ragnarok nordicum* has never been completely described. The only published morphological data are contained in a brief description of M1/-M3/ by Jepsen (1930) and Archibald's (1982) description of a possibly associated /M1-2 and a dentary fragment with damaged and heavily worn teeth (both from the lower Tullock Formation).

Description of the anterior lower dentition (/C, /P1-3) is based on a nearly complete but highly fractured dentary, UCMP 134797 (Figure 28). Two mental foramina are present, one below the posterior root of /P3 and another below the anterior root of /P2. The latter foramen is poorly preserved because of the highly fractured condition of the anterior part of the mandible. The canine is large, somewhat conical, with its apex extending much higher than the cusps of the premolars. The canine root is recurved beneath /P1. A single-rooted, slightly procumbent /P1 is separated from the canine by a small diastem. The /P1 has a small posterior basal cusp and a very small anterior basal cuspule, which is shifted slightly lingual to the midline of the tooth. The root of /P1 is the size of the posterior root of the double-rooted /P2. /P2 has a large central cusp which is higher than that of /P1, and a weakly developed talonid with a posterior basal cusp. An anterior basal cusp is not evident, but the enamel is slightly swollen in this region. The /P3 is broken off at root level and was not recovered; however, it had two roots. A diastem is not present between any of the premolars.

Description of remaining parts of the lower dentition is based on a much larger sample of associated (Table 50) and isolated teeth. The /P4 of *Ragnarok nordicum* has a massive protoconid and a large, low metaconid. The paraconid is large but is lower than the metaconid and is shifted lingually to the midline of the tooth (Figure 29). A protolophid is well developed on many specimens. The posterior labial cingulum is well developed, and a weaker anterior labial cingulum is present. These cingula are not connected across the protoconid. The talonid is well developed, with a large cusp in the position of the hypoconid (Figures 28, 29). The /P4 is widest at the trigonid-talonid junction.

The /P4 of *Ragnarok* is clearly larger than any of the other "condylarths" found in the upper Hell Creek Formation except *Protungulatum gorgun*. In comparison to *P. gorgun*, the /P4 of *Ragnarok* is longer and much wider, its protoconid is more bulbous, the posterior labial cingulum is better developed, and the paraconid and metaconid are relatively larger. Also, unlike *Protungulatum gorgun*, *Ragnarok* lacks premolar diastems.

Lower molars of *Ragnarok nordicum* have low trigonids with bulbous cusps that have greatly swollen lateral walls (see Figures 28, 29). The metaconid is slightly higher and larger than the protoconid. The paraconid is closely appressed to the metaconid and is much smaller than the other trigonid cusps. Also, the paraconid is shifted labially in relation to a line extending through the metaconid and entoconid. A small but distinct paralophid is present.

The talonid cusps are low but robust. The hypoconid is the largest cusp on /M1-2, with the entoconid and hypoconulid of subequal size. The hypoconulid is the largest cusp on /M3 and is usually inclined anteriorly. The hypoconid is slightly smaller, and the entoconid is anteroposteriorly elongated into a ridge. The labial cingulum is strong on all lower molars, and can be continuous from the paraconid posteriorly to the hypoconulid. The lingual cingulum is well developed at the talonid notch, but is rarely present elsewhere on the lingual side of the teeth. Distinct ectostylids are present on all lower molars, and the enamel in the hypoflexid region is commonly crenulated or rugose, especially on /M2.

In comparison to *Protungulatum gorgun*, the lower molars of *Ragnarok* are larger; have lower trigonids and more bulbous trigonid and talonid cusps; the trigonid walls of *Ragnarok* are greatly expanded; and the paraconid is closely appressed to the metaconid. The paraconid does not (or only slightly) projects anteriorly in *Ragnarok*.

Measurements of the lower dentition of *R. nordicum* from McGuire Creek are given in Tables 49 and 50.

A maxillary fragment with a nearly complete post-canine dentition (UCMP 134694, Figure 30B-D, Table 51) forms the basis for the description of the anterior premolars of *Ragnarok nordicum*. P1-P3/ were separated from the maxillary fragment during discovery. Unfortunately, the poor condition of the maxillary fragment is such that the anterior premolars cannot be remounted to ascertain their orientation or spacing.

The P1/ is single-rooted, but the crown is missing. Based on the size of the root, it was roughly comparable in size to the lower /P1. P2/ has two roots and a large anteroposteriorly elongated central cusp (Figure 30C). A prominent "metacrista" extends posteriorly to end in a small metastylar lobe. A small but distinct metastyle is present on a well developed labial cingulum. P3/ has a paracone and is more molariform than P2/, but still lacks any trace of a metacone (Figure 30D). The paracone is large, bulbous, and has a distinct metacrista. The metastylar lobe is large, with a prominent metastyle. The parastylar lobe is much smaller and has a small parastyle. The labial cingulum is continuous across the paracone. The protoconal lobe is missing, but the tooth narrows rather abruptly lingual to the paracone. If a protocone was present, it was small.

As with P3/, P4/ lacks a metacone but is more molariform because of the lingual extension of the protoconal lobe, which gives the tooth a transverse appearance (Figure

Table 52. Measurements of associated and isolated upper molars of *Ragnarok nordicum* from McGuire Creek localities and Mantua Lentil[a]

|  | McGuire Creek *R. nordicum* | Mantua Lentil *R. nordicum* |
|---|---|---|
| **M1/ Length** | | |
| Number | 3 | 3 |
| Observed Range | 4.99-5.50 | 4.80-5.27 |
| Mean | 5.17 | 5.09 |
| **M1/ W-A** | | |
| Number | 3 | 3 |
| Observed Range | 6.46-6.73 | 5.57-6.47 |
| Mean | 6.58 | 6.03 |
| **M1/ W-P** | | |
| Number | 2 | 3 |
| Observed Range | 6.89-6.91 | 6.02-6.89 |
| Mean | 6.90 | 6.38 |
| **M2/ Length** | | |
| Number | 3 | 4 |
| Observed Range | 5.67-5.92 | 5.64-6.03 |
| Mean | 5.81 | 5.87 |
| **M2/ W-A** | | |
| Number | 3 | 4 |
| Observed Range | 6.92-8.13 | 7.13-7.96 |
| Mean | 7.70 | 7.66 |
| **M2/ W-P** | | |
| Number | 3 | 3 |
| Observed Range | 7.21-8.47 | 7.62-8.48 |
| Mean | 7.98 | 8.13 |
| **M3/ Length** | | |
| Number | 4 | 3 |
| Observed Range | 4.68-5.94 | 4.21-5.70 |
| Mean | 5.27 | 4.99 |
| **M3/ W-A** | | |
| Number | 4 | 3 |
| Observed Range | 6.28-7.01 | 5.40-6.35 |
| Mean | 6.66 | 5.87 |
| **M3/ W-P** | | |
| Number | 4 | 3 |
| Observed Range | 5.57-6.37 | 5.00-5.60 |
| Mean | 5.99 | 5.38 |

a. Measurements from Mantua Lentil were taken from casts of PU 13285 (M1-M2/, M3/ fragment), PU 16870 (M1-3/), PU 21011 (P4/-M2/), and PU 16854 (P4/-M3/).

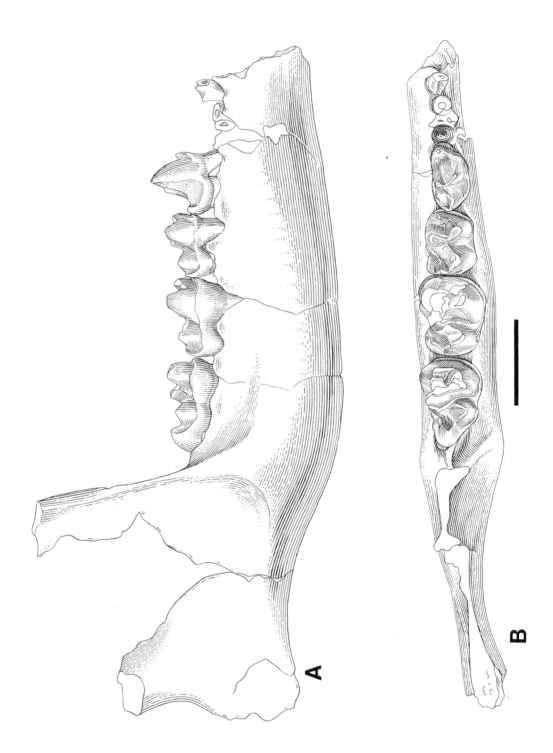

A

B

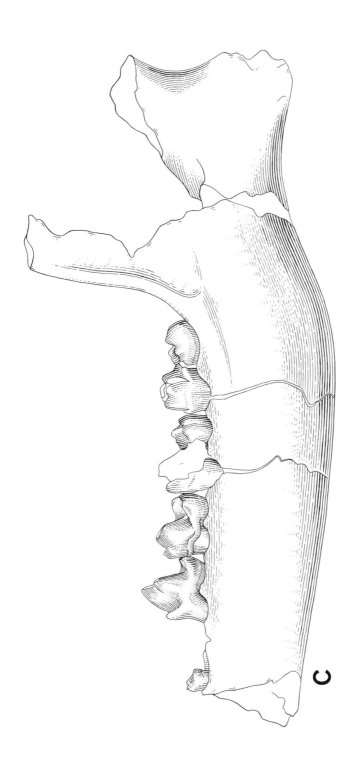

C

FIGURE 29. *Ragnarok nordicum* Jepsen. Right dentary with /P4, /M1-3, UCMP 132458, locality V86031: (A) labial view, (B) occlusal view, (C) lingual view. Scale bar = 5 mm.

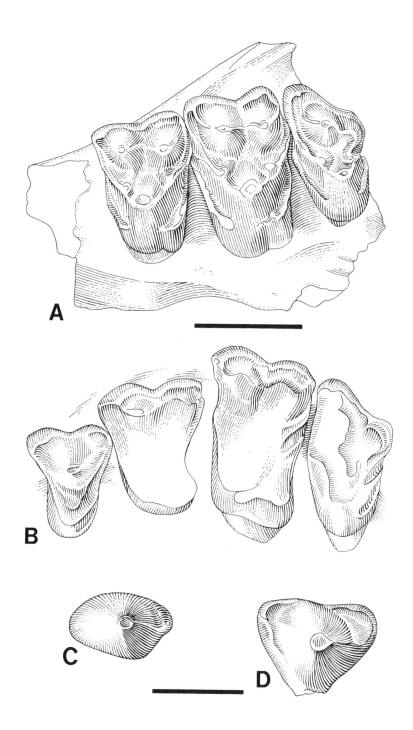

FIGURE 30. *Ragnarok nordicum* Jepsen. (A) Left maxillary fragment with M1-3/, UCMP 134693, locality V87072, occlusal view. (B) Left maxillary fragment with P2-3, M1-3, UCMP 134694, locality V87072, occlusal view of P4/, M1-3/. Scale bar = 6 mm. Occlusal views of (C) P2/, and (D) P3/. Scale bar = 3 mm.

30B). The protocone is large, and a weak pre- and postcingula are present. The paracone is larger than the protocone and has a metacristid extending posteriorly to a large metastylar lobe with a distinct metastyle. The parastylar lobe is smaller and has a small parastyle. The labial cingulum is continuous across the paracone, and an ectoflexus is well developed.

The upper molars of *Ragnarok nordicum* were described by Jepsen (1930), and only a few additional comments are necessary based on fossils from McGuire Creek. Upper molars have massive protocones and large conules lacking or with weak internal wings, except M3/, which usually has a strong premetaconule wing (Figure 30A). The M1/ has a weak paracrista and a distinct parastyle. A small mesostyle is usually present on M3/. Posterior lingual cingula are well developed, with distinct but small hypocones on M1-2/ (Figure 30A-B). M3/ lacks a hypocone. The lingual cingula are not continuous across the protocone in upper molars, with the exception of a single M2/.

In comparison to *Protungulatum gorgun*, the upper molars of *Ragnarok nordicum* have relatively more bulbous paracones and metacones. Also, their protocones are larger and much broader antero-posteriorly, giving the upper molars a more quadrate outline.

Measurements of associated and isolated upper molars and premolars of *Ragnarok nordicum* from McGuire Creek are given in Tables 51 and 52.

Family PERIPTYCHIDAE Cope, 1882c
Subfamily ANISONCHINAE Osborn and Earle, 1895
*Mimatuta* Van Valen, 1978
*Mimatuta morgoth* Van Valen, 1978
Tables 53-56; Figure 31

*Mimatuta morgoth* Van Valen, 1978, p. 62.

*Type.* UMVP 1560, right maxilla with M2/ (Van Valen, 1978, pl. 7, fig. 5).

*Type locality.* Harbicht Hill, Hell Creek Formation, Montana.

*Referred specimens.* Dentary with /P4, /M2-3, UCMP 132454, from loc. V86031. Dentary with /P4-/M2, UCMP 132340, from loc. V87029. One /P4, 1 M3/ from loc. V87037. Two /P4's from loc. V87038. One M3/ from loc. V87049. One /P4, 1 M2/ from loc. V87072. One M1/ from loc. V87077. Dentary with /P4-/M3, UCMP 134590, from loc. V87088. Two M1/'s, 2 M3/'s from loc. V87098. One M3/ from loc. V87101. Dentary with /C, /P1-4, /M1-3, UCMP 134589, from loc. V87123. Two /P4's, 1 M3/ from loc. V87151.

*Localities.* UCMP locs. V86031, V87029, V87037, V87038, V87049, V87072, V87077, V87088, V87098, V87101, V87123, and V87151.

*Distribution.* Upper Hell Creek Formation and lower Tullock formations, Montana (both Puercan).

*Description.* Only a few isolated upper molars from McGuire Creek are referable to *Mimatuta morgoth*. Archibald (1982) has adequately described the upper molars of *M. morgoth*, and only a few additional comments are required.

Archibald's (1982) admittedly tentative identification of the tooth site of UCMP 116518 is incorrect (this conclusion is based on comparison of UCMP 116518 to speci-

Table 53.  Measurements of a cast of the type of *Mimatuta morgoth*, an isolated M2/ (UMVP 1560) from Harbicht Hill (V71203), and measurements of isolated upper molars of *M. morgoth* from McGuire Creek localities

| Locality | Specimen | Tooth Site | Length | W-A | W-P | A | A/W-A |
|----------|----------|-----------|--------|-----|-----|---|-------|
| V87098 | 134640 | M1/ | 4.05 | 5.26 | 5.31 | 2.30 | .44 |
| V87098 | 134634 | M1/ | 4.57 | 5.63 | 5.79 | 2.18 | .39 |
| V87072 | 133446 | M2/ | 4.77 | 6.16 | 6.54 | 2.75 | .45 |
| V71203 | UMVP 1560 | M2/ | 4.76 | 5.96 | 6.24 | 2.42 | .41 |
| V87037 | 132620 | M3/ | 4.21 | 5.20 | 4.75 | 1.89 | .36 |
| V87049 | 133148 | M3/ | 4.22 | 4.85 | 4.37 | 1.85 | .38 |
| V87098 | 134639 | M3/ | 3.85 | 5.27 | 4.60 | 2.03 | .39 |
| V87098 | 133853 | M3/ | 3.94 | 4.88 | 4.53 | 1.80 | .37 |
| V87101 | 134574 | M3/ | 4.19 | 5.38 | 4.91 | 2.27 | .42 |
| V87151 | 132227 | M3/ | 4.32 | 5.45 | 5.20 | 2.16 | .40 |

mens of *Mimatuta morgoth* from McGuire Creek). It is an M1/, not an M2/. Otherwise Archibald's description of upper molars of *M. morgoth* applies equally well to fossils from McGuire Creek.

The diagnosis of *Mimatuta* included the statement that the protocone was central in position (i.e. shifted labially) (Van Valen, 1978). Using a small sample (n=3), Archibald (1982) proposed that this relative labial shift of the protocone can be used to distinguish *M. morgoth* from the similar-sized species *M. minuial* and *Protungulatum donnae*. Recent work by Luo (1989, 1991), employing statistical analysis of the type of *Mimatuta morgoth*, an isolated M2/ from Harbicht Hill, and a large sample of upper molars of arctocyonids from Bug Creek Anthills, supports use of this character in separating *Protungulatum donnae* and *Mimatuta morgoth*. Analysis of the sample of *M. morgoth* from McGuire Creek (Table 53, n=9) also supports the use of this character in distinguishing species of "condylarths." The relative labial shift of the protocone is expressed in the A/W-A ratio (i.e., length of the lingual protoconal slope divided by the anterior width of the tooth), which is different in each of the three species. Examination of A/W-A ratios of fossils studied by Archibald (1982), the type of *M. morgoth*, and the McGuire Creek sample of the three species shows that *M. minuial* has higher and *Protungulatum donnae* lower A/W-A ratios than *Mimatuta morgoth* (Table 54). Therefore, this character appears to have diagnostic utility, as Archibald (1982) suggested.

Lower molars of *Protungulatum donnae* can be separated from those of *Mimatuta* by the labial shift of the paraconids in the latter (Archibald, 1982). But in contrast to the upper dentition, lower molars of *M. morgoth* are difficult to distinguish from those of *M. minuial*. The diagnosis of each species only recognizes differences in the size of the /P4 metaconid as the diagnostic character. The /P4 metaconid is "rather strong" in *M.*

Table 54. A/W-A ratios of upper molars of *Mimatuta morgoth, M. minuial,* and *Protungulatum donnae* from McGuire Creek, Harbicht Hill V71203, Worm Coulee 1 V74111, and Bug Creek Anthills V70210 (=V65127)[a]

|  |  | *P. donnae* | *M. morgoth* | *M. minuial* |
|---|---|---|---|---|
| M1/ | Number | 4 | 3 | |
| A/W-A | Ratio | .34-.37 | .39-.47 | |
| M2/ | Number | 10 | 3 | 2 |
| A/W-A | Ratio | .32-.38 | .41-.45 | .48 |
| M3/ | Number | 3 | 7 | |
| A/W-A | Ratio | .27-.30 | .31-.42 | |

a. Sources for A/W-A ratios: *M. morgoth,* V74111, Archibald, 1982, table 49; V71203, Type M2/ UMVP 1560, and McGuire Creek sample, table 53. *P. donnae,* V70201, Archibald, 1982, appendix 4; McGuire Creek sample, Table 40. *M. minuial,* V74111, Archibald, 1982, Table 51; McGuire Creek sample, Table 57.

*minuial* and "rather weak" in *M. morgoth* (see Van Valen, 1978). Isolated and associated (with molars in dentaries) /P4's of *Mimatuta* from McGuire Creek can be separated into two distinct morphotypes, those with strong and those with weak metaconids. Five dentary fragments from McGuire Creek referable to *Mimatuta,* on the basis of the position of the molar paraconids, have the /P4 in place. Four of the five dentaries have weakly developed /P4 metaconids and are referred to *M. morgoth* (Table 55). The other dentary has a much more robust /P4 metaconid and is referred to *M. minuial* (Table 56). Comparison of molar morphology of dentaries referred to the respective species provides no clearly distinctive characters. Archibald (1982) suggested that the relative width of the molars that he tentatively referred to *M. minuial* (n=2) were greater than those of *M. morgoth.* Dentaries referred to each species from McGuire Creek on the basis of the development of the /P4 metaconid, show that molars of *M. minuial* and *M. morgoth* are approximately equal in relative width (Table 56) and cannot be distinguished on this basis. Casts of dentaries of *M. minuial* from Mantua Lentil, the type locality of the species, were measured and are included in Table 56. Comparison of molar talonid widths of *M. minuial* from Mantua Lentil and *M. minuial* and *M. morgoth* from McGuire Creek show that relative talonid width has questionable diagnostic value. Therefore, isolated lower molars of *Mimatuta* from McGuire Creek cannot be referred to either *M. minuial* or *M. morgoth* with confidence, and are assigned to *Mimatuta* species indeterminate.

Archibald (1982) has adequately described the lower dentition of *M. morgoth* except for the anterior parts of the mandible and its associated dentition. Based on more complete material from McGuire Creek, description of this part of the dentition is given below.

A well preserved dentary of *M. morgoth* with /C, /P1-4, /M1-3, (UCMP 134589,

Table 55. Measurements of associated lower premolars and molars of *Mimatuta morgoth* from McGuire Creek localities

| Locality | Specimen | Site | Tooth Length | W-Tri | W-Tal |
|----------|----------|------|--------|-------|-------|
| V87123 | 134589 | /C | 2.08 | 1.60 (width) | |
| | 134589 | /P1 | 1.54 | 1.10 (width) | |
| | 134589 | /P2 | 2.50 | 1.34 (width) | |
| | 134589 | /P3 | 3.16 | 1.71 (width) | |
| | 134589 | /P4 | 3.44 | 2.39 (width) | |
| | 134589 | /M1 | 3.92 | 2.98 | 2.92 |
| | 134589 | /M2 | 4.09 | 3.31 | 3.03 |
| | 134589 | /M3 | 5.21 | 2.90 | 2.54 |
| V87088 | 134590 | /P4 | 3.80 | 2.41 (width) | |
| | 134590 | /M1 | 4.05 | 2.98 | 3.00 |
| | 134590 | /M2 | 4.56 | 3.62 | 3.49 |
| | 134590 | /M3 | 5.43 | 3.28 | 2.54 |
| V86031 | 132454 | /P4 | 3.45 | 2.31 (width) | |
| | 132454 | /M2 | 3.97 | 3.20 | 2.81 |
| | 132454 | /M3 | 4.90 | 2.85 | 2.33 |
| V87029 | 132340 | /P4 | 3.75 | 2.27 (width) | |
| | 132340 | /M1 | 3.92 | — | 2.76 |
| | 132340 | /M2 | 4.25 | 3.44 | 3.17 |

Figure 31, Table 55) shows that the canine has a vertically oriented, robust crown whose apex extends above /P1 and /P2 and equals the height of the central cusp of /P3. The canine has a weakly developed anterior ridge along the midline of the crown, and its root is strongly recurved posteriorly, extending beneath the root of /P1. The /P1 is slightly procumbent, and its central cusp is lower than that of /P2.

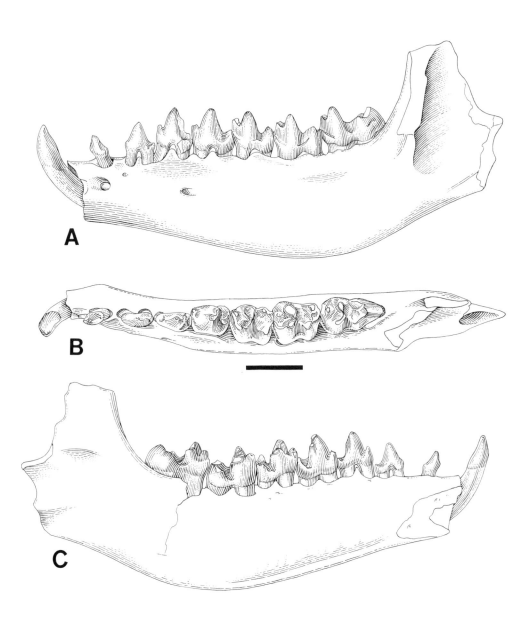

FIGURE 31. *Mimatuta morgoth* Van Valen. Left dentary with /C, /P1-4, /M1-3, UCMP 134589, locality V87123: (A) labial view, (B) occlusal view, (C) lingual view. Scale bar = 4 mm.

Table 56. Measurements of associated lower premolars and molars of *Mimatuta morgoth* from McGuire Creek localities and measurements of *M. minuial* from V87072 (McGuire Creek) and Mantua Lentil

| | *M. morgoth*<br>McGuire Creek<br>N=4 for /P4, /M2<br>N=3 for /M1, /M3 | *M. minuial*<br>V87072<br>N=1 | *M. minuial*<br>Mantua Lentil<br>N=3 |
|---|---|---|---|
| /P4 Length | | | |
| Observed Range | 3.44-3.80 | 3.82 | 3.37-3.58 |
| Mean | 3.61 | 3.82 | 3.47 |
| /P4 Width | | | |
| Observed Range | 2.27-2.41 | 2.10 | 2.42-2.59 |
| Mean | 2.35 | 2.10 | 2.49 |
| /M1 Length | | | |
| Observed Range | 3.92-4.05 | 4.13 | 3.63-3.77 |
| Mean | 3.96 | 4.13 | 3.71 |
| /M1 W-Tri | | | |
| Observed Range | 2.98-2.98 | 2.93 | 2.66-2.91 |
| Mean | 2.98 | 2.93 | 2.78 |
| /M1 W-Tal | | | |
| Observed Range | 2.76-3.00 | 2.88 | 2.70-2.88 |
| Mean | 2.89 | 2.88 | 2.77 |
| /M1 W-Tri/W/Tal | | | |
| Observed Range | .99-1.02 | 1.02 | .99-1.02 |
| Mean | 1.01 | 1.02 | 1.01 |
| /M1 Length/W-Tal | | | |
| Observed Range | 1.34-1.35 | 1.43 | 1.29-1.39 |
| Mean | 1.35 | 1.43 | 1.34 |
| /M2 Length | | | |
| Observed Range | 3.97-4.56 | 4.37 | 3.76-4.13 |
| Mean | 4.22 | 4.37 | 3.97 |
| /M2 W-Tri | | | |
| Observed Range | 3.20-3.62 | 3.50 | 3.15-3.47 |
| Mean | 3.39 | 3.50 | 3.29 |
| /M2 W-Tal | | | |
| Observed Range | 2.81-3.49 | 3.17 | 2.94-3.29 |
| Mean | 3.13 | 3.17 | 3.15 |
| /M2 W-Tri/W-Tal | | | |
| Observed Range | 1.04-1.14 | 1.10 | 1.01-1.07 |
| Mean | 1.09 | 1.10 | 1.04 |
| /M2 Length/W-Tal | | | |
| Observed Range | 1.31-1.41 | 1.38 | 1.22-1.28 |
| Mean | 1.35 | 1.38 | 1.26 |

| | *M. morgoth* McGuire Creek N=4 for /P4, /M2 N=3 for /M1, /M3 | *M. minuial* V87072 N=1 | *M. minuial* Mantua Lentil N=3 |
|---|---|---|---|
| /M3 Length | | | |
| Observed Range | 4.90-5.43 | 5.02 | 4.40-4.71 |
| Mean | 5.18 | 5.02 | 4.57 |
| /M3 W-Tri | | | |
| Observed Range | 2.85-3.28 | 3.13 | 2.76-3.14 |
| Mean | 3.01 | 3.13 | 2.95 |
| /M3 W-Tal | | | |
| Observed Range | 2.33-2.54 | 2.50 | 2.38-2.70 |
| Mean | 2.47 | 2.50 | 2.54 |
| /M3 W-Tri/W-Tal | | | |
| Observed Range | 1.14-1.29 | 1.25 | 1.16-1.17 |
| Mean | 1.22 | 1.25 | 1.16 |
| /M3 Length/W-Tal | | | |
| Observed Range | 2.05-2.14 | 2.01 | 1.74-1.85 |
| Mean | 2.10 | 2.01 | 1.80 |

*Mimatuta minuial* Van Valen, 1978
Tables 54, 56-57; Figure 32

*Mimatuta minuial* Van Valen, 1978, p. 62.

*Type.* PU 14211, left maxilla with P3-4/, M1-3/ (Van Valen, 1978, pl. 7, fig. 4).

*Type locality.* Mantua Lentil, Polecat Bench Formation (Fort Union Formation), Wyoming.

*Referred specimens.* One /P4 from loc. V87036. One /P4 from loc. V87038. Dentary with /P4-/M3, UCMP 134695; 1 /P4 from loc. V87072. Maxilla with M2/, UCMP 132308, from loc. V87091.

*Localities.* UCMP locs. V87036, V87038, V87072, and V87091.

*Distribution.* Upper Hell Creek and lower Tullock formations, Montana; Polecat Bench Formation (Fort Union Formation), Wyoming (all Puercan).

*Description.* The only specimen from McGuire Creek containing part of the upper dentition that confidently can be referred to *Mimatuta minuial* is a maxillary fragment containing a damaged M2/ (UCMP 132308). The M2/ is enamel-less (except for part of the trigon basin) and the parastylar lobe is missing. In spite of this damage, it is clear that the protocone of UCMP 132308 is shifted labially to a great degree, producing a high A/W-A ratio (Table 57), more so than in *M. morgoth* (Table 53). Also, remnants of the internal conular wings are preserved, and they appear to have been well developed, more so than in specimens of *M. morgoth* from McGuire Creek.

As discussed in the description of *Mimatuta morgoth*, the strong /P4 metaconid of *M. minuial* appears to provide the basis for separation of the two species. Casts of den-

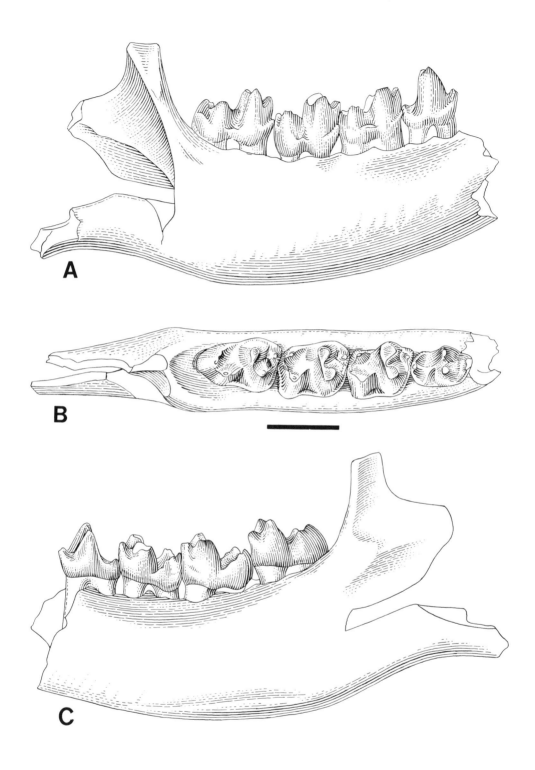

FIGURE 32. *Mimatuta minuial* Van Valen. Right dentary fragment with /P4, /M1-3, UCMP 134695, locality V87072: (A) labial view, (B) occlusal view, (C) lingual view. Scale bar = 4 mm.

Table 57. Measurements of UCMP 132308, an isolated M2/ of *Mimatuta minuial* from locality V87091

| Length | W-A | W-P | A | A/W-A |
|--------|-----|-----|-----|-------|
| 3.90 | 5.78[a] | 6.09 | 2.79 | .48[a] |

a: Approximate measurement.

taries of the hypodigm from Mantua Lentil, the type locality of *M. minuial*, all have strong /P4 metaconids. The /P4 metaconid of UCMP 134695 from McGuire Creek is robust, and nearly equals the protoconid in height and size (Figure 32). Isolated /P4's from McGuire Creek have metaconids that are smaller and lower than their respective protoconids, but these are still much better developed than metaconids of *M. morgoth*.

As discussed in the description of *M. morgoth*, lower molars of *M. minuial* and *M. morgoth* apparently are not distinguishable unless they are associated with the /P4. Isolated lower molars and dentary fragments of *Mimatuta* without /P4's are referred to *Mimatuta* species indeterminate.

### *Mimatuta* sp. indet.

*Referred specimens.* One /M2 from loc. V84151. One /M1, 1 /M3 trigonid from loc. V84193. Dentary with /M2 talonid, /M3, UCMP 132301, from loc. V84194. One /M1 from loc. V87030. One /M2 from loc. V87033. Dentary with /M1-2, /M3 trigonid, UCMP 132474, from loc. V87034. One /M1 from loc. V87035. Associated /P3-4, /M1-3, UCMP 134557; 1 /M2, 1 /M3 from loc. V87037. One /M1, 1 /M3 from loc. V87038. One /M2 from loc. V87049. One /M3 from loc. V87067. One /M2 from loc. V87071. One /M1, 1 /M3 from loc. V87072. One /M2 from loc. V87073. One /M1 from loc. V87074. One /M1, 1 /M2 from loc. V87077. Dentary with /M1-2, UCMP 132808; 1 /M2, 1 /M3 from loc. V87084. Dentary with /M1-2, UCMP 132317, from loc. V87095. One /M3 from loc. V87101. One /M1 from loc. V87151.

*Localities.* UCMP locs. V84151, V84193, V84194, V87030, V87033, V87034, V87035, V87037, V87038, V87049, V87067, V87071, V87072, V87073, V87074, V87077, V87084, V87095, V87101, and V87151.

*Description.* Includes dentary fragments and isolated lower molars that probably belong to either *Mimatuta morgoth* and *M. minuial* but are too fragmentary or incomplete to be referable to either species with confidence. The number of identifiable teeth of *M. morgoth* (upper molars, isolated /P4's, and dentaries with /P4's) from McGuire Creek greatly outnumber those referred to *M. minuial*. Therefore, it is likely that the majority of fossils placed in this category belong to *M. morgoth*.

# Appendix 2
# Vertebrate Faunal Lists

A partial vertebrate faunal list for each local fauna (Plates 1-4), its proposed biochronologic age, and the mammalian component for each UCMP fossil locality are given below. The surface exposure of all UCMP fossil localities was prospected. If additional collection techniques were employed for a locality, these are listed under the applicable local fauna. An asterisk (*) denotes that a fossil locality is plotted within measured sections on Plates 2-4. Up-Up-the-Creek and Brown-Grey Channel local faunas and section locations are plotted on Plate 1. Precise locality information is on file at the University of California Museum of Paleontology, Berkeley, and is available to qualified investigators. All localities yield isolated dinosaur teeth unless otherwise indicated.

Taxa whose presence in Puercan (Pu0-Pu1 interval zones) local faunas at McGuire Creek may be entirely due to reworking are: *Meniscoessus robustus*, *Essonodon browni*, *Cimolodon nitidus*, *Pediomys hatcheri*, *P. krejcii*, *P. florencae*, *P. elegans*, *Pediomys* sp. indet., *Alphadon "wilsoni"*, *A. rhaister*, *A. marshi*, *Alphadon* sp. indet., *Glasbius twitchelli*, *Didelphodon vorax*, *Gypsonictops illuminatus*, and *Batodon tenuis*.

## LOWER TULLOCK FORMATION
*Jacks Channel Local Fauna*: Puercan (Pu1?); UCMP locs. V84190, V88036; 375 lbs. of sediment were screenwashed from both localities combined. Dinosaur teeth and bone fragments absent.
    MULTITUBERCULATA
*Mesodma* sp.
*Stygimys* sp. indet.
    EUTHERIA
*Procerberus* sp. indet.
    "CONDYLARTHRA"
?*Mimatuta* sp.
Locality Faunal List
    V84190 Jacks Ridge N 4: *Stygimys* sp. indet.
    V88036 Luck O Hutch: *Mesodma* sp.; *Procerberus* sp. indet.; ?*Mimatuta* sp.

HELL CREEK FORMATION
*K-Mark Channel Local Fauna:* Lancian; UCMP loc. V85092 only; 300 lbs. of sediment were screenwashed.
   *V85092 K-Mark 2: dinosaur teeth abundant, dinosaur bone fragments present.
   MULTITUBERCULATA
*Meniscoessus robustus*
   MARSUPIALIA
*Pediomys florencae*
*Didelphodon vorax*
   EUTHERIA
*Gypsonictops illuminatus*

Other Lancian Localities
   *V84182 Jacks Bay 1: abundant dinosaur teeth.
   *V87044 Matt's Dino Quarry: associated hadrosaur elements.

*Z-line Channel Local Fauna:* Puercan (Pu1); UCMP locs. V84193, V84194; 300 lbs. of sediment were screenwashed from both localities combined. Dinosaur teeth and bone fragments absent.
   MULTITUBERCULATA
*Mesodma* sp.
   MARSUPIALIA
*Peradectes* cf. *P. pusillus*
   EUTHERIA
*Procerberus formicarum*
   "CONDYLARTHRA"
*Mimatuta* sp. indet.
Locality Faunal List
   V84193   Z-Line Quarry: *Mesodma* sp.; *Procerberus formicarum*; *Mimatuta* sp. indet.
   *V84194   Z-Line E Quarry: *Mesodma* sp.; *Peradectes* cf. *P. pusillus*; *Procerberus formicarum*; *Mimatuta* sp. indet.

*Black Spring Coulee Channel Local Fauna:* Puercan (Pu1?); UCMP locs. V87030, V87031, V87033, V87123, V87124, V88044; 500 lbs. of sediment were screenwashed from locality V87030. Dinosaur teeth and bone fragments present, large dinosaur bones abundant at locality V87030.
   MULTITUBERCULATA
*Mesodma* sp.
*Stygimys kuszmauli*
*Cimexomys gratus*
*Cimexomys minor*
*Catopsalis* sp. indet.
*Meniscoessus robustus*

EUTHERIA

*Procerberus formicarum*

"CONDYLARTHRA"

*Protungulatum donnae*

*Protungulatum gorgun*

*Mimatuta morgoth*

*Mimatuta* sp. indet.

*Ragnarok nordicum*

Locality Faunal List.

*V87030   Black Spring Coulee S: *Mesodma* sp.; *Stygimys kuszmauli; Cimexomys gratus; C. minor; Meniscoessus robustus; Procerberus formicarum; Mimatuta* sp. indet.; *Ragnarok nordicum.*

V87031   Black Spring Coulee N 1: *Protungulatum donnae.*

V87033   Black Spring Coulee N 3: *Catopsalis* sp. indet.; *Mimatuta* sp. indet.; *Ragnarok nordicum.*

V87123   Come Alive Condylarth: *Mimatuta morgoth.*

V87124   Condylarth Flats: *Stygimys kuszmauli; Protungulatum gorgun; Ragnarok nordicum.*

V88044   Late Celebration: *Ragnarok nordicum.*

*Up-Up-The-Creek Local Fauna*: Puercan (Pu1); UCMP locs. V84151, V87034, V87035, V87036, V87037, V87038; 2750 lbs. of sediment were screenwashed from localities V87035-38 combined. Dinosaur teeth and bone fragments present.

MULTITUBERCULATA

*Mesodma* sp.

*Stygimys kuszmauli*

*Cimexomys gratus*

*Cimexomys minor*

*Catopsalis alexanderi*

*Catopsalis* sp. indet.

*Meniscoessus robustus*

MARSUPIALIA

*Peradectes* cf. *P. pusillus*

*Pediomys hatcheri*

*Pediomys krejcii*

*Pediomys florencae*

*Pediomys elegans*

*Pediomys* sp. indet.

*Alphadon "wilsoni"*

*Alphadon rhaister*

*Alphadon* sp. indet.

*Glasbius twitchelli*

*Didelphodon vorax*

EUTHERIA
*Procerberus formicarum*
*Gypsonictops illuminatus*
*Batodon tenuis*
　"CONDYLARTHRA"
*Protungulatum donnae*
*Protungulatum gorgun*
*Mimatuta morgoth*
*Mimatuta minuial*
*Mimatuta* sp. indet.
*Oxyprimus erikseni*
*Ragnarok nordicum*
Locality Faunal List.

　V84151　Up-the-Creek: *Meniscoessus robustus; Mimatuta* sp. indet.; *Ragnarok nordicum.*

　V87034　Up-Up-the-Creek 1: *Mesodma* sp.; *Stygimys kuszmauli; Catopsalis alexanderi; Meniscoessus robustus; Didelphodon vorax; Protungulatum donnae; P. gorgun; Mimatuta* sp. indet.; *Oxyprimus erikseni; Ragnarok nordicum.*

　V87035　Up-Up-the-Creek 1A: *Mesodma* sp.; *Stygimys kuszmauli; Cimexomys gratus; Catopsalis* sp. indet.; *Meniscoessus robustus; Peradectes* cf. *P. pusillus; Pediomys hatcheri; P. florencae; Alphadon* sp. indet.; *Glasbius twitchelli; Didelphodon vorax; Procerberus formicarum; Gypsonictops illuminatus; Protungulatum donnae; P. gorgun; Mimatuta* sp. indet.; *Oxyprimus erikseni; Ragnarok nordicum.*

　V87036　Up-Up-the-Creek 1B: *Mesodma* sp.; *Stygimys kuszmauli; Pediomys krejcii; Alphadon rhaister; Procerberus formicarum; Protungulatum donnae; Mimatuta minuial; Oxyprimus erikseni; Ragnarok nordicum.*

　V87037　Up-Up-the-Creek 2: *Mesodma* sp.; *Stygimys kuszmauli; Cimexomys gratus; C. minor; Catopsalis alexanderi; Meniscoessus robustus; Peradectes* cf. *P. pusillus; Pediomys hatcheri; P. krejcii; P. elegans; Pediomys* sp. indet.; *Alphadon "wilsoni"; Alphadon* sp. indet.; *Glasbius twitchelli; Didelphodon vorax; Procerberus formicarum; Protungulatum donnae; P. gorgun; Mimatuta morgoth; Mimatuta* sp. indet.; *Ragnarok nordicum.*

　V87038　Up-Up-the-Creek 3: *Mesodma* sp.; *Stygimys kuszmauli; Cimexomys gratus; C. minor; Meniscoessus robustus; Peradectes* cf. *P. pusillus; Pediomys florencae; P. elegans; Pediomys* sp. indet.; *Alphadon* sp. indet.; *Glasbius twitchelli; Procerberus formicarum; Gypsonictops illuminatus; Batodon tenuis; Protungulatum donnae; P. gorgun; Mimatuta morgoth; M. minuial; Mimatuta* sp. indet.; *Oxyprimus erikseni; Ragnarok nordicum.*

*Brown-Grey Channel Local Fauna*: Puercan (Pu1); UCMP locs. V87040, V87070, V87071, V87072; 1175 lbs. of sediment screenwashed from locality V87072; V87072 was hand-quarried; V87071 and V87072 both include anthill samples. Large pieces of dinosaur bone and dinosaur teeth abundant.

MULTITUBERCULATA

*Mesodma* sp.

*Stygimys kuszmauli*

*Cimexomys gratus*

*Cimexomys minor*

*Catopsalis joyneri*

*Catopsalis* sp. indet.

*Meniscoessus robustus*

*Cimolodon nitidus*

MARSUPIALIA

*Peradectes* cf. *P. pusillus*

*Pediomys hatcheri*

*Pediomys krejcii*

*Pediomys florencae*

*Pediomys elegans*

*Pediomys* sp. indet.

*Alphadon "wilsoni"*

*Alphadon rhaister*

*Alphadon* sp. indet.

*Alphadon marshi*

*Glasbius twitchelli*

*Didelphodon vorax*

EUTHERIA

*Procerberus formicarum*

*Gypsonictops illuminatus*

"CONDYLARTHRA"

*Protungulatum donnae*

*Protungulatum gorgun*

*Mimatuta morgoth*

*Mimatuta minuial*

*Mimatuta* sp. indet.

*Oxyprimus erikseni*

*Ragnarok nordicum*

Locality Faunal List

V87040   BC Bone-anza: *Stygimys kuszmauli; Pediomys florencae; Protungulatum donnae.*

V87070   Tedrow Quarry B: *Mesodma* sp.; *Stygimys kuszmauli; Catopsalis* sp. indet.; *Didelphodon vorax.*

*V87071 Tedrow Quarry C: *Mesodma* sp.; *Stygimys kuszmauli; Cimexomys gratus; C. minor; Catopsalis* sp. indet.; *Catopsalis joyneri; Meniscoessus robustus; Pediomys elegans; Pediomys* sp. indet.; *Procerberus formicarum; Gypsonictops illuminatus; Protungulatum gorgun; Mimatuta* sp. indet.; *Oxyprimus erikseni.*

*V87072  Tedrow Quarry D: *Mesodma* sp.; *Stygimys kuszmauli; Cimexomys gratus; C. minor; Meniscoessus robustus; Cimolodon nitidus; Peradectes* cf. *P. pusillus; Pediomys hatcheri; P. krejcii; P. flo-*

*rencae; P. elegans; Alphadon "wilsoni"; Alphadon rhaister; A. marshi; Alphadon* sp. indet.; *Glasbius twitchelli; Didelphodon vorax; Procerberus formicarum; Gypsonictops illuminatus; Protungulatum donnae; Mimatuta morgoth; M. minuial; Mimatuta* sp. indet.; *Oxyprimus erikseni; Ragnarok nordicum.*

*Lower Tedrow Channel Local Fauna*: Puercan (Pu1?); UCMP locality V87152 only; quarry sample. Associated ceratopsian skeletal elements present.

    \*V87152  Tedrow Dinosaur Quarry
    MULTITUBERCULATA
*Stygimys kuszmauli*
    EUTHERIA
    "CONDYLARTHRA"
*Ragnarok nordicum*

*Upper Tedrow Channel Local Fauna*: Puercan (Pu1); UCMP locality V88037 only.

    V88037  Upper Tedrow
    MULTITUBERCULATA
*Meniscoessus robustus*

*Shiprock Local Fauna*: Puercan (Pu1); UCMP localities V87073, V87074, V87077, V87153; 50 lbs. of sediment were screenwashed from locality V87153; V87073, V87074, V87077 include anthill samples. Dinosaur teeth and bone fragments present.

    MULTITUBERCULATA
*Mesodma* sp.
*Stygimys kuszmauli*
*Cimexomys gratus*
*Cimexomys minor*
*Catopsalis joyneri*
*Meniscoessus robustus*
    MARSUPIALIA
*Peradectes* cf. *P. pusillus*
*Pediomys krejcii*
*Pediomys elegans*
*Alphadon "wilsoni"*
*Alphadon rhaister*
*Glasbius twitchelli*
*Didelphodon vorax*
    EUTHERIA
*Procerberus formicarum*
*Gypsonictops illuminatus*
*Batodon tenuis*

"CONDYLARTHRA"
*Protungulatum donnae*
*Protungulatum gorgun*
*Mimatuta morgoth*
*Mimatuta* sp. indet.
*Oxyprimus erikseni*
*Ragnarok nordicum*
Locality Faunal List

V87073  Grass Patch: *Stygimys kuszamuli; Alphadon "wilsoni"; Didelphodon vorax; Procerberus formicarum; Mimatuta* sp. indet.; *Ragnarok nordicum.*

V87074  Shiprock: *Mesodma* sp.; *Stygimys kuszmauli; Cimexomys gratus; C. minor; Catopsalis* sp. indet.; *Meniscoessus robustus; Peradectes* cf. *P. pusillus; Pediomys krejcii; P. elegans; Alphadon "wilsoni"; A. rhaister; Glasbius twitchelli; Procerberus formicarum; Gypsonictops illuminatus; Batodon tenuis; Mimatuta* sp. indet.; *Oxyprimus erikseni; Ragnarok nordicum.*

*V87077 North Edge: *Mesodma* sp.; *Stygimys kuszmauli; Cimexomys gratus; C. minor; Protungulatum gorgun; Mimatuta morgoth; Mimatuta* sp. indet.; *Ragnarok nordicum.*

V87153  Bad Mouth Turtle: *Mesodma* sp.; *Stygimys kuszmauli; Catopsalis joyneri; Meniscoessus robustus; Procerberus formicarum; Gypsonictops illuminatus; Oxyprimus erikseni; Ragnarok nordicum.*

*Second Level Channel Local Fauna*: Puercan (Pu1?); UCMP localities V86031, V87029, V87050, V87051, V87052, V87078, V87091, V87095, V87101, V87114, V87117, V87119, V87151; 325 lbs. of sediment were screenwashed from localities V87114, V87101, V87091, and V87052 combined; V87051, V87052, V87078, V87151 include anthill samples. Dinosaur teeth and bone fragments present.

MULTITUBERCULATA
*Mesodma* sp.
*Stygimys kuszmauli*
*Cimexomys gratus*
*Cimexomys minor*
*Catopsalis joyneri*
*Meniscoessus robustus*
*Essonodon browni*

MARSUPIALIA
*Pediomys hatcheri*
*Pediomys krejcii*
*Pediomys florencae*
*Pediomys elegans*
*Alphadon "wilsoni"*
*Alphadon rhaister*
*Didelphodon vorax*

EUTHERIA
*Procerberus formicarum*
*Gypsonictops illuminatus*
*Batodon tenuis*
    "CONDYLARTHRA"
*Protungulatum donnae*
*Protungulatum gorgun*
*Mimatuta morgoth*
*Mimatuta minuial*
*Mimatuta* sp. indet.
*Oxyprimus erikseni*
*Ragnarok nordicum*
Locality Faunal List

V86031  Jaw Breaker: *Stygimys kuszmauli; Meniscoessus robustus; Protungulatum donnae; Mimatuta morgoth; Oxyprimus erikseni; Ragnarok nordicum.*

V87029  Eagle Nest Channel 2: *Didelphodon vorax; Protungulatum donnae; P. gorgun; Mimatuta morgoth; Oxyprimus erikseni; Ragnarok nordicum.*

V87050  Eagle High Extension: *Ragnarok nordicum.*

V87051  Eagle High: *Mesodma* sp.; *Stygimys kuszmauli; Cimexomys gratus; C. minor; Pediomys elegans; Procerberus formicarum; Gypsonictops illuminatus; Oxyprimus erikseni.*

V87052  Eagle South: *Mesodma* sp.; *Stygimys kuszmauli; Cimexomys gratus; Pediomys elegans; Alphadon rhaister; Oxyprimus erikseni.*

*V87078  Three Buttes 1: *Stygimys kuszmauli; Didelphodon vorax; Ragnarok nordicum.*

V87091  Eagle Nest Southeast: *Mesodma* sp.; *Cimexomys minor; Mimatuta minuial.*

*V87095  Rattlesnake Nest: *Mimatuta* sp. indet.; *Oxyprimus erikseni.*

V87101  Second Level: *Mesodma* sp.; *Stygimys kuszmauli; Cimexomys gratus; Alphadon "wilsoni"; Gypsonictops illuminatus; Protungulatum donnae; Mimatuta morgoth; Mimatuta* sp. indet.

*V87114  Second Level South: *Mesodma* sp.; *Stygimys kuszmauli; Cimexomys gratus; Procerberus formicarum.*

V87117  Eagle Nest Ridge West: *Ragnarok nordicum.*

V87119  Lone Juniper Tree North: *Protungulatum gorgun.*

V87151  Juniper Tree West: *Mesodma* sp.; *Stygimys kuszmauli; Cimexomys gratus; Catopsalis joyneri; Meniscoessus robustus; Essonodon browni; Pediomys hatcheri; P. krejcii; P. florencae; Didelphodon vorax; Procerberus formicarum; Gypsonictops illuminatus; Batodon tenuis; Protungulatum donnae; Mimatuta morgoth; Mimatuta* sp. indet.; *Oxyprimus erikseni; Ragnarok nordicum.*

*Little Roundtop Channel Local Fauna*: Puercan (Pu0); UCMP localities V87028, V87094, V87098, V87115, V88038; 375 lbs. of sediment were screenwashed from localities V87028, V87098, V87115 combined; V87098 includes anthill sample. Dinosaur teeth and bone fragments present.

MULTITUBERCULATA
*Mesodma* sp.
*Stygimys kuszmauli*
*Cimexomys gratus*
*Cimexomys minor*
*Meniscoessus robustus*
  MARSUPIALIA
*Pediomys hatcheri*
*Pediomys elegans*
*Alphadon rhaister*
*Alphadon* sp. indet.
*Didelphodon vorax*
  EUTHERIA
*Procerberus formicarum*
*Gypsonictops illuminatus*
  "CONDYLARTHRA"
*Protungulatum donnae*
*Protungulatum gorgun*
*Mimatuta morgoth*
*Oxyprimus erikseni*
*Ragnarok nordicum*
Locality Faunal List

V87028 Eagle Nest Channel 1: *Mesodma* sp.; *Stygimys kuszmauli; Cimexomys gratus; Pediomys elegans; Procerberus formicarum; Protungulatum gorgun; Oxyprimus erikseni.*

*V87094 Pick Pocket: Mammals not recovered.

*V87098 Little Roundtop: *Mesodma* sp.; *Stygimys kuszmauli; Cimexomys gratus; C. minor; Meniscoessus robustus; Pediomys hatcheri; P. elegans; Alphadon rhaister; Alphadon* sp. indet.; *Didelphodon vorax; Procerberus formicarum; Gypsonictops illuminatus; Protungulatum donnae; P. gorgun; Mimatuta morgoth; Oxyprimus erikseni; Ragnarok nordicum.*

*V87115 Lower Level SW: *Stygimys kuszmauli; Protungulatum gorgun.*

*V88038 Leafbranch Level: *Stygimys kuszmauli; Protungulatum donnae.*

*The Swamp Local Fauna*: Puercan (Pu1?); UCMP localities V85085, V85086, V86093. Associated and articulated skeletons of champsosaurs, turtles, and crocodiles. Dinosaur teeth and bones not present.

*V85085 Lone Tree Coulee S 1.

*V85086 Lone Tree Coulee S 2.

*V86093 Lone Tree Coulee S 4.

Isolated Localities (Puercan?)

*V87066 Coulee Seven-Four: *Mesodma* sp.

*V88042 Yellow Butte Pocket: Mammals not recovered.

"Three Buttes Local Fauna"

   V87082  Three Buttes 5: *Protungulatum donnae; Ragnarok nordicum.*

   V87083  Three Buttes 6: *Didelphodon vorax.*

   V87084  Three Buttes 7: *Mesodma* sp.; *Stygimys kuszmauli; Meniscoessus robustus; Protungulatum donnae; P. gorgun; Mimatuta* sp. indet.; *Oxyprimus erikseni.*

   V87086  Three Buttes 9: *Stygimys kuszmauli.*

   V87088  Three Buttes 10: *Protungulatum gorgun; Mimatuta morgoth; Ragnarok nordicum.*

# Literature Cited

Alexopoulos, J.S., R.A.F. Grieve, and P.B. Robertson
1988    Microscopic lamellar deformation features in quartz: Discriminative character-
        istics of shock generated varieties. Geology 16:796-799.
Alvarez, L.W., W. Alvarez, F. Asaro, and H.V. Michel
1980    Extraterrestrial cause for the Cretaceous-Tertiary extinction. Science
        208:1095-1108.
Alvarez, W.
1986    Toward a theory of impact crises. Eos (Amer. Geophys. Union Trans.) 67(35):649-658.
Alvarez, W., L.W. Alvarez, F. Asaro, and H.V. Michel
1982    Current status of the impact theory for the terminal Cretaceous extinction, pp.
        305-315 in L.T. Silver and P.H. Schultze, eds., Geological implications of impacts
        of large asteroids and comets on the earth. Geol. Soc. Amer. Spec. Paper 190.
1984    The end of the Cretaceous: sharp boundary or gradual transition? Science
        223:1183-1186.
Ameghino, F.
1887    Enumeración sistemática de las especies de mamíferos fósiles coleccionados por
        Carlos Ameghino en los terrenos eocenos de la Patagonia austral y depositados
        en el Museo de La Plata. Bol. Mus. de La Plata 1:1-26.
1889    Contribución al conocimento de los mamíferos fósiles de la Republica
        Argentina, obra escrita bajo los auspicios de al Academia Nacional de Ciencias
        de la República Argentina para presentarla a la Exposición Universal de Paris de
        1889. Actas Acad. Cienc. Córdoba 6:1-1027.
1890    Los plagiaulacideos argentinos y sus relaciones zoológicas, geológicas y geográ-
        ficas. Bol. Instituto Geográfico Argentino 9:143-201.
1894    Enumération synoptique des especies de mammiféres fossiles des formations
        éocenes de Patagonie. Bol. Acad. Cienc. Córdoba, Buenos Aires 13:259-452.
1901    Notices préliminaires sur de ongules nouveaux de terrains crétacés de
        Patagonie. Bol. Acad. Nac. Cienc. Córdoba, Buenos Aires 16:349-426.
Archibald, J.D.
1981    The earliest known Paleocene fauna and its implications for the Cretaceous-Tertiary
        boundary. Nature 208:650-652.

1982    A study of Mammalia and geology across the Cretaceous-Tertiary boundary in Garfield County, Montana. Univ. Calif. Publ. Geol. Sci. 122:1-286.

1984    Bug Creek Anthills (BCA), Montana: faunal evidence for Cretaceous age and non-catastrophic extinctions. Geol. Soc. Amer. Abst. Prog. 16:432.

1986    Comment on "Sedimentology, stratigraphy, and extinctions during the Cretaceous-Paleogene transition at Bug Creek, Montana." Geology 14:892-893.

1987a   Latest Cretaceous and early Tertiary mammalian biochronology/biostratigraphy in the Western Interior. Geol. Soc. Amer. Abst. Prog. 19:258.

1987b   The Bugcreekian Land Mammal Age: a reassessment. Jour. Vert. Paleo. 7(3):10A.

1987c   Stepwise and non-catastrophic late Cretaceous terrestrial extinctions in the Western Interior of North America: testing observations in the context of an historical science. Mem. de la Soc. Geol. de France, n.s. 150:45-52.

Archibald, J.D., and L.J. Bryant

1990    Differential Cretaceous/Tertiary extinctions of nonmarine vertebrates; evidence from northeastern Montana, pp. 549-562 in V.L. Sharpton and P.D. Ward, eds., Global Catastrophes in Earth History: an interdisciplinary conference on Impacts, Volcanism, and Mass Mortality. Geol. Soc. Amer. Spec. Paper 247.

Archibald, J.D., R.F. Butler, E.H. Lindsay, W.A. Clemens, and L.W. Dingus

1982    Upper Cretaceous-Paleocene biostratigraphy and magnetostratigraphy, Hell Creek and Tullock formations, northeastern Montana. Geology 10:153-159.

Archibald, J.D., and W.A. Clemens

1984    Mammal evolution near the Cretaceous-Tertiary Boundary. pp. 339-371, in W.A. Berggren and J.A. Van Couvering, eds., Catastrophes and Earth History. Princeton University Press, Princeton, New Jersey.

Archibald, J.D., W.A. Clemens, P.D. Gingerich, D.W. Krause, E.H. Lindsay, and K.D. Rose

1987    First North American Land Mammal Ages of the Cenozoic Era, pp. 24-76

Archibald, J.D., and D.L. Lofgren

1990    Mammalian Zonation near the Cretaceous-Tertiary Boundary, pp. 31-50 in K.D. Rose and T. Bown, eds., Dawn of the Age of Mammals in the northern part of the Rocky Mountain Interior, North America. Geol. Soc. Amer. Spec. Paper 243.

Argast, S., J.O. Farlow, R.M. Gabet, and D.L. Brinkman

1987    Transport induced abrasion of fossil reptilian teeth: implications for the existance of Tertiary dinosaurs in the Hell Creek Formation, Montana. Geology 15:927-930.

Badgley, C.

1986    Counting individuals in mammalian fossil assemblages from fluvial environments. Palaios 1:328-338.

Behrensmeyer, A.K.

1982    Time resolution in fluvial vertebrate assemblages. Paleobiology 8:211-227.

Belt, E.S., R.M. Flores, P.D. Warwick, K.M. Conway, K.R. Johnson, and R.S. Waskowitz

1984    Relationship of fluviodeltaic facies to coal deposition in the Lower Fort Union Formation (Paleocene), south-western North Dakota. Special Publications International Association Sedimentologists. 7:177-195.

Berggren, W.A.
1964    The Maastrichtian, Danian, and Montian stages and the Cretaceous-Tertiary boundary. Stockholm Cont. Geology 11:103-176.

Berry, W.B.N.
1984    The Cretaceous-Tertiary boundary—the ideal geologic time scale boundary? Newslet. Strat. 13:143-155.

Bohor, B.F., E.E. Foord, P.J. Modreski, and D.M. Triplehorn
1984    Mineralogic evidence for an impact event at the Cretaceous-Tertiary boundary. Science 224:867-869.

Bohor, B.F., P.J. Modreski, and E.E. Foord
1987a    Shocked quartz in the Cretaceous-Tertiary Boundary Clays: Evidence for a Global Distribution. Science 236:705-709.

Bohor, B.F., D.M. Triplehorn, D.J. Nichols, and H.T. Millard, Jr.
1987b    Dinosaurs, spherules, and the "magic" layer: a new K-T boundary clay site in Wyoming. Geology 15:896-899.

Brown, B.
1907    The Hell Creek beds of the upper Cretaceous of Montana. Bull. Amer. Mus. Nat. Hist. 23:823-845.

Brown, R.
1952    Tertiary strata in eastern Montana and western North and South Dakota. Billings Geol. Soc. Guidebook 3:89-92.

Bryant, L.J.
1989    Non-dinosaurian lower vertebrates across the Cretaceous-Tertiary Boundary in northeastern Montana. Univ. Calif. Publ. Geol. Sci. 134:1-107.

Bryant, L.J., J.H. Hutchison, W.A. Clemens, and J.D. Archibald
1986    Diversity changes in lower vertebrates (non-dinosaurian) across the Cretaceous-Tertiary boundary in northeastern Montana. Geol. Soc. Amer. Abst. Prog. 18:522.

Butler, R.D.
1980    Stratigraphy, sedimentology, and depositional environments of the Hell Creek Formation (late Cretaceous) and adjacent strata, Glendive area, Montana. Ph.D. dissertation, University of North Dakota, Grand Forks. 398 pp.

Calvert, W.R.
1912    Geology of certain lignite fields in eastern Montana. U.S. Geol. Surv. Bull. 471:187-201.

Carlson, C.G., and S.B. Anderson
1966    Sedimentary and tectonic history of North Dakota part of Williston Basin. Amer. Assoc. Pet. Geol. Bull. 49:1833-1846.

Carter, N.L., C.B. Officer, C.A. Chesner, and W.I. Rose
1986    Dynamic deformation of volcanic ejecta from the Toba caldera: Possible relevance to Cretaceous/Tertiary boundary phenomena. Geology 14:380-383.

Cherven, V.B., and A.F. Jacob
1985    Evolution of Paleogene depositional systems, Williston Basin, in response to global sea level changes, pp. 127-170 in R.M. Flores and S.S Kaplan, eds., Cenozoic Paleogeography of the West Central United States. Rocky Mt. Symp. 3, Rocky Mt. Sec., Soc. Econ. Paleo. Min., Denver.

Cifelli, R.L.
1990    Cretaceous mammals of southern Utah. I. Marsupials from the Kaiparowits Formation (Judithian). Jour. Vert. Paleo. 10:295-319.

Clark, J., J.R. Beerbower, and K.K. Kietzke
1967    Oligocene sedimentation, paleoecology, and paleoclimatology in the Big Badlands of South Dakota. Field. Geol. Mem. 5:1-158.

Clemens, W.A.
1964    Fossil mammals of the type Lance Formation, Wyoming: Part I. Introduction and Multituberculata. Univ. Calif. Publ. Geol. Sci. 48:1-105.
1966    Fossil mammals of the type Lance Formation, Wyoming: Part II. Marsupialia. Univ. Calif. Publ. Geol. Sci. 62:1-122.
1968    A mandible of *Didelphodon vorax* (Marsupialia, Mammalia). Los Angeles Co. Mus. Cont. Sci. 133:1-11.
1973    Fossil mammals of the type Lance Formation, Wyoming: Part III. Eutheria and Summary. Univ. Calif. Publ. Geol. Sci. 94:1-102.
1982    Patterns of extinction and survival of the terrestrial biota during the Cretaceous/Tertiary transition, pp. 407-413 in L.T Silver and P.H. Schultze, eds., Geological implications of impacts of large asteriods and comets on the earth. Geol. Soc. Amer. Spec. Paper 190.
1983    Mammalian evolution during the Cretaceous-Tertiary transition; Evidence for gradual, non-catastrophic patterns of biotic change. Acta Palaeo. Polonica 28:55-61.

Clemens, W.A., and J.D. Archibald
1980    Evolution of terrestrial faunas during the Cretaceous-Tertiary transition. pp. 67-74, in Ecosystemes Continentaux du Mesozoique. Mem. de la Soc. Geol. de France, n.s. 59.

Clemens, W.A., J.D. Archibald, and L.J. Hickey
1981    Out with a whimper not a bang. Paleobiology 7:293-298.

Cobban, W.A.
1958    Late Cretaceous fossil zones of the Powder River Basin, Wyoming and Montana. Wyoming Geol. Assoc. Guidebook 13:114-119.

Collier, A.J., and M. Knechtel
1939    The coal resources of McCone County, Montana. U.S. Geol. Surv. Bull. 905:1-80.

Cope, E.D.
1882a   A second genus of Eocene Plagiaulacidae. Amer. Nat. 16:416-417.
1882b   Mammalia in the Laramie Formation. Amer. Nat. 16:830-831.
1882c   The Periptychidae. Amer. Nat. 16:832-833.
1884    The Tertiary Marsupialia. Amer. Nat. 18:686-697.

Crochet, J.-Y.

1979    Diversité systématique des Didelphidae (Marsupialia) Européens Tertiares. Geobios 12:365-378.

Crocket, J.H., C.B. Officer, F.C. Wezel, and G.D. Johnson

1988    Distribution of noble metals across the Cretaceous-Tertiary boundary at Gubbio, Italy: Iridium variation as a constant on the duration and nature of Cretaceous/Tertiary boundary events. Geology 16:77-80.

Cuvier, G.

1817    Le regne animal. Paris, Deterville 1: xxxvii + 540 pp.

Dingus, L.W.

1983    A stratigraphic review and analysis for selected marine and terrestrial sections spanning the Cretaceous-Tertiary boundary. Ph.D. dissertation, University of California, Berkeley. 156 pp.

1984    Effects of stratigraphic completeness on interpretations of extinction rates across the Cretaceous-Tertiary boundary. Paleobiology 10:420-438.

Dorf, E.

1942    Upper Cretaceous floras of the Rocky Mountain region. II: Flora of the Lance Formation at its type locality, Niobrara County, Wyoming. Carnegie Inst. Wash. Publ. 508:79-159.

Dyer, B.D., N.N. Lyalikova, D. Murray, M. Doyle, G.M. Kolesov, and W.E. Krumbein

1989    Role for microorganisms in the formation of iridium anomalies. Geology 17:1036-1039.

Eaton, J.G., J.I. Kirkland, and K. Doi

1989    Evidence of reworked Cretaceous fossils and their bearing on the existence of Tertiary dinosaurs. Palaios 4:281-286.

Ely, J.C., and J.K. Rigby, Jr.

1989    Abrasion features on vertebrate teeth as a subjective index of transport distance and exposure to traction/saltation loads. Geol. Soc. Amer. Abst. Prog. 21:A97.

Estes, R., and P. Berberian

1970    Paleoecology of a Late Cretaceous vertebrate community from Montana. Breviora 343:1-35.

Fastovsky, D.E.

1986    Paleoenvironments of vertebrate-bearing strata at the Cretaceous-Paleogene boundary in northeastern Montana and southwestern North Dakota. Ph.D. dissertation, University of Wisconsin, Madison. 301 pp.

1987    Paleoenvironments of vertebrate-bearing strata during the Cretaceous-Paleogene transition, eastern Montana and western North Dakota. Palaios 2:282-295.

Fastovsky, D.E., and R.H. Dott

1986    Sedimentology, stratigraphy, and extinctions during the Cretaceous-Paleogene transition at Bug Creek, Montana. Geology 14:279-282.

Fastovsky, D.E., and K. McSweeney

1987    Paleosols spanning the Cretaceous-Paleogene transition, eastern Montana and western North Dakota. Geol. Soc. Amer. Bull. 99:66-77.

Felder, P.J., W.M. Felder, and R.G. Bromley

1980    The type area of Maastrichtian stage, pp. 118-137 in T. Birkelund and R.G. Bromley, eds., The Upper Cretaceous and Danian of Northwestern Europe, Int. Geol. Cong. 26, Guidebook 69A.

Fox, R.C.

1978    Upper Cretaceous terrestrial vertebrate stratigraphy of the Gobi Desert (Mongolian People's Republic) and western North America, pp. 577-594 in C.R. Stelck and B.D.E. Chatterton, eds., Western and Arctic Canadian Biostratigraphy. Geol. Assoc. Can. Spec. Paper 18.

1987    Patterns of mammalian evolution towards the end of the Cretaceous, Saskatchewan, Canada, pp. 7-11 in Short Papers, Fourth Symposium on Mesozoic Terrestrial Ecosystems. Occ. Pap. Tyrrell Mus. Palaeo. 3.

1989    The Wounded Knee Local Fauna and Mammalian Evolution near the Cretaceous-Tertiary Boundary, Saskatchewan, Canada. Palaeontographica Abt. A, 208:11-59.

1990    The succession of Paleocene mammals in western Canada, pp. 51-70 in T.M. Bown and K.D. Rose, eds., Dawn of the Age of Mammals in the northern part of the Rocky Mountain Interior, North America. Geol. Soc. Amer. Spec. Paper 243.

Fox, R.C., and B.G. Naylor

1986    A new species of *Didelphodon* Marsh (Marsupialia) from the Upper Cretaceous of Alberta, Canada: Paleobiology and Phylogeny. N. Jb. Geol. Palaont. Abh. 172:357-380.

Fox, S.K., Jr., and R.K. Olsson

1969    Danian planktonic foraminifera from the Cannonball Formation in North Dakota. Jour. Paleo. 43:1397-1404.

Frye, C.I.

1969    Stratigraphy of the Hell Creek Formation in North Dakota: North Dakota Geol. Surv. Bull. 54:1-65.

Galbreath, E.C.

1959    Collecting fossils from harvester ant mounds. Trans. Kansas Acad. Sci. 62:173-174.

Giebel, C.G.A.

1855    Die Saugethiere in zoologischer, anatomischer, und palaeontologischer Beziehung um fassend dargestellt. Leipzig, Ambrosius Abel. 1108 pp.

Gill, J.R., and W.A. Cobban.

1973    Stratigraphy and geologic history of the Montana Group and equivalent rocks, Montana, Wyoming, and North and South Dakota. U.S. Geol. Surv. Prof. Paper 776:1-37.

Gill, T.

1872    Arrangement of the families of mammals with analytical tables. Smithsonian Misc. Coll. 11:1-98.

Granger, W., and G.G. Simpson

1929    A revision of Tertiary Multituberculata. Bull. Amer. Mus. Nat. Hist. 56:601-676.

Gregory, W.K., and G.G. Simpson

1926    Cretaceous mammal skulls from Mongolia. Amer. Mus. Nov. 225:1-20.

Hahn, G., and R. Hahn

1983    Multituberculata. Fossilium Catalogus 1. Animalia 127:1-409.

Hansen, H.J., R. Gwozdz, R.G. Bromley, K.L. Rasmussen, E.W. Vogensen, and K.R. Pedersen

1986    Cretaceous-Tertiary boundary spherules from Denmark, New Zealand, and Spain. Bull. Geol. Soc. Denmark 35:75-82.

Harland, W.B., A.V. Cox, P.G. Llewellyn, G.A.C. Pickton, A.G. Smith, and R. Walters

1982    A Geologic Time Scale: Cambridge University Press, Cambridge, England. 131 pp.

Hatcher, J.B.

1896    Some localities for Laramie mammals and horned dinosaurs. Amer. Nat. 30:112-120.

Hotton, C.

1984    Palynofloral changes across the Cretaceous-Tertiary boundary in east- central Montana. U.S.A. Abst. Inter. Paly. Conf. 6:66.

1988    Palynology of the Cretaceous-Tertiary Boundary in central Montana, U.S.A., and its implications for Extraterrestrial Impact. Ph.D. dissertation, University of California, Davis. 610 pp.

Hut, P., W. Alvarez, W.P. Elder, T. Hansen, E.G. Kauffman, G. Keller, E.M. Shoemaker, and P.R. Weissman

1987    Comet showers as a cause of mass extinctions. Nature 329:118-126.

Hutchison, J.H., and J.D. Archibald

1986    Diversity of turtles across the Cretaceous-Tertiary boundary in northeastern Montana. Palaeogeogr. Palaeoclimatol. Palaeoecol. 55:1-22.

Huxley, T.H.

1880    On the application of the laws of evolution to the arrangement of the Vertebrata and more particularly of the Mammalia. Proc. Zool. Soc. London, pp. 649-662.

Illiger, C.

1811    Prodromus systematis mammalium et avium additis terminis zoographicis utri-udque classis. C. Salfeld, Berlin. 301 pp.

Jeletsky, J.A.

1960    Youngest marine rocks in the Western Interior of North America and the age of the *Triceratops* beds; with remarks on comparable dinosaur-bearing beds outside North America. Proc. 21st Int. Geol. Cong., part 5:25-40.

1962    The allegedly Danian dinosaur-bearing rocks of the globe and the problem of the Mesozoic-Cenozoic boundary. Jour. Paleo. 36:1005-1018.

Jensen, F.S., and H.D. Varnes

1964    Geology of the Fort Peck area, Garfield, McCone, and Valley counties, Montana. U.S. Geol. Surv. Prof. Paper 414-F:1-49.

Jepsen, G.L.

1930    Stratigraphy and paleontology of the Paleocene of northeastern Park County, Wyoming. Proc. Amer. Phil. Soc. 69:463-528.

1940    Paleocene faunas of the Polecat Bench Formation, Park County, Wyoming. Part I. Proc. Amer. Phil. Soc. 83:217-341.

Jerzykiewicz, T., and A.R. Sweet
1986 The Cretaceous-Tertiary boundary in the central Alberta foothills. I: Stratigraphy. Can. Jour. Earth Sci. 23:1356-1374.

Johanson, Z.
1991 A revision of *Alphadon marshi* Simpson, 1927, and *Alphadon wilsoni* Lillegraven, 1969 (Marsupialia). Jour. Vert. Paleo. 11(3):38A-39A.

Johnson, K.R., D.J. Nichols, M. Attrep, Jr., and C.J. Orth
1989 High-resolution leaf-fossil record spanning the Cretaceous-Tertiary boundary. Nature 340:708-711.

Johnston, P.A.
1980 First record of Mesozoic mammals from Saskatchewan. Can. Jour. Earth Sci.17:512-519.

Johnston, P.A., and R.C. Fox
1984 Paleocene and Late Cretaceous mammals from Saskatchewan, Canada. Palaeontographica Abt. A, 186:163-222.

Keller, G.
1989 Extended period of extinctions across the Cretaceous/Tertiary boundary in planktonic foraminifera of continental-shelf sections: Implications for impact and volcanism theories. Geol. Soc. Amer. Bull. 101:1408-1419.

Kielan-Jaworowska, Z., and R.E. Sloan
1979 *Catopsalis* (Multituberculata) from Asia and North America and the problem of taeniolabidid dispersal in the Late Cretaceous. Acta Palaeo. Polonica 24:187-197.

Krishtalka, L., and R.K. Stucky
1983 Paleocene and Eocene Marsupials of North America. Ann. Carnegie Mus. 52:229-263.

Kuslys, M., and U. Krahenbulh
1983 Noble metals in Cretaceous/Tertiary sediments from El Kef. Radiochimica Acta 34:139-141.

Leffingwell, H.A.
1970 Palynology of the Lance (Late Cretaceous) and Fort Union (Paleocene) formations of the type Lance area. Geol. Soc. Amer. Spec. Paper 127:1-64.

Lerbekmo, J.F., and R.M. St. Louis
1986 The terminal Cretaceous iridium anomaly in the Red Deer Valley, Alberta, Canada. Can. Jour. Earth Sci. 23:120-124.

Lerbekmo, J.F., A.R. Sweet, and R.M. St. Louis
1987 The relationship between the iridium anomaly and palynological floral at three Cretaceous-Tertiary boundary localities in western Canada. Geol. Soc. Amer. Bull. 99:325-330.

Lillegraven, J.A.
1969 Latest Cretaceous mammals of upper part of Edmonton Formation of Alberta, Canada, and review of marsupial-placental dichotomy of mammalian evolution. Univ. Kansas Paleo. Cont. 50 (Vertebrata 12):1-122.

Lillegraven, J.A., and M.C. McKenna

1986    Fossil mammals from the "Mesaverde" Formation (Late Cretaceous, Judithian) of the Bighorn and Wind River Basins, Wyoming, with definitions of Late Cretaceous North American Land-Mammal "Ages." Amer. Mus. Nov. 2840:1-68.

Linnaeus, C.

1758    Systema naturae per regna tria naturae, secundum classes, ordines, genera, species, cum characteribus, differentiis, synonymis, locis. Editio decima reformata, vol. 1. Stockholm, Laurentii Salvii. ii + 824 pp.

1766    Systema naturae per regna tria naturae secundum classes, ordines, genera, species, cum characteribus differentiis, synonymis, locis. Editio decima reformata 1(7). Stockholm, Laurentii Salvii. 532 pp.

Lofgren, D.L.

1992    Upper premolar configuration of *Didelphodon vorax* (Mammalia, Marsupialia, Stagodontidae). Jour. Paleo. 66:162-164.

Lofgren, D.L., and C. Hotton

1988    Palynologically defined Cretaceous Bug Creek Facies channel, upper Hell Creek Formation, McCone Co., N.E. Montana. Geol. Soc. Amer. Abst. Prog. 20:428.

1991a   The Bug Creek problem and the K-T transition at McGuire Creek, Mt. Geol. Soc. Amer. Abst. Prog. 23:A359.

1991b   The K-T transition at McGuire Creek Montana. Jour. Vert. Paleo. 11(3):43A-44A.

Lofgren, D.L., C. Hotton, and A.C. Runkel

1990    Reworking of Cretaceous Dinosaurs into Paleocene Channel Deposits, Upper Hell Creek Formation, Montana. Geology 18:874-877.

Lull, R.S.

1915    The mammals and horned dinosaurs of the Lance Formation, Niobrara County, Wyoming. Amer. Jour. Sci., ser. 4, 40:319-348.

Luo, Z.

1989    Structure of the petrosals of Multituberculata (Mammalia) and morphology of the molars of early arctocyonids (Condylarthra, Mammalia). Ph.D. dissertation, University of California, Berkeley. 426 pp.

1991    Variability of dental morphology and the relationships of the earliest arctocyonid species. Jour. Vert. Paleo. 11:452-471.

Lupton, C., D. Gabriel, and R.M. West

1980    Paleobiology and depositional setting of a Late Cretaceous vertebrate locality, Hell Creek Formation, McCone County, Montana. Univ. Wyoming Cont. Geol. 18:117-126.

McKenna, M.C.

1965    Collecting microvertebrate fossils by washing and screening, pp. 193-203 in B. Kummer D. and Raup, eds., Handbook of Paleontological Techniques. W.H. Freeman. San Francisco.

1975    Toward a phylogenetic classification of the Mammalia, pp. 21-46 in W.P. Luckett and F.S. Szalay, eds., Phylogeny of the Primates. Plenum Press, New York.

Macdonald, L.J.

1972    Monroe Creek (Early Miocene) microfossils from the Wounded Knee area, South Dakota. South Dakota Geol. Surv. Rep. Invest. 105:1-43.

MacLeod, N., and G. Keller

1991a   Hiatus distributions and mass extinctions at the Cretaceous/Tertiary boundary. Geology 19:497-501.

1991b   How complete are Cretaceous/Tertiary boundary sections? A chronostratigraphic estimate based on graphic correlation. Geol. Assoc. Amer. Bull. 103:1439-1457.

Marsh, O.C.

1880    Notice of Jurassic mammals representing two new orders. Amer. Jour. Sci., ser. 3, 20:235-239.

1889a   Discovery of Cretaceous Mammalia, part 1. Amer. Jour. Sci., ser. 3, 38:81-92.

1889b   Discovery of Cretaceous Mammalia, part 2. Amer. Jour. Sci., ser. 3, 38:177-180.

1892    Discovery of Cretaceous Mammalia, part 3. Amer. Jour. Sci., ser. 3, 48:249-262.

Marshall, L.G., J.A. Case, and M.O. Woodburne

1990    Phylogenetic relationships of the families of marsupials, pp. 433-505 in H.H. Genoways, ed., Current Mammalogy, vol. 2. Plenum Press, New York.

Matthew, W.D.

1937    Paleocene faunas of the San Juan Basin, New Mexico. Trans. Amer. Phil. Soc. 30:1-510.

Matthew, W.D., and W. Granger

1921    New genera of Paleocene mammals. Amer. Mus. Nov. 13:1-17.

Middleton, M.D.

1982    A new species and additional material of *Catopsalis* (Mammalia, Multituberculata) from the Western Interior of North America. Jour. Paleo. 56:1197-1206.

1983    Early Paleocene vertebrates of the Denver Basin, Colorado. Ph.D. dissertation, University of Colorado, Boulder. 404 pp.

Montanari, A.

1986    Spherules from the Cretaceous/Tertiary boundary clay at Gubbio, Italy: The problem of outcrop contamination. Geology 14:1024-1026.

Montanari, A., R.L. Hay, W. Alvarez, F. Asaro, H.V. Michel, and L.W. Alvarez

1983    Spheroids at the Cretaceous-Tertiary boundary are altered impact droplets of basaltic composition. Geology 11:668-671.

Moore, W.L.

1976    The stratigraphy and environments of deposition of the Cretaceous Hell Creek Formation (reconnaissance) and the Paleocene Ludlow Formation (detailed), southwestern North Dakota. North Dakota Geol. Surv. Rep. Invest. 56:1-40.

Mossup, G.D., and P.D. Flach

1983    Deep channel sedimentation in the Lower Cretaceous McMurray Formation, Athabasca Oil Sands, Alberta. Sedimentology 30:493-509.

Murray, A.

1866    The Geographic Distribution of Mammals. Day and Son, London. 420 pp.

Naslund, H.R., C.B. Officer, and G.D. Johnson

1986      Microspherules in Upper Cretaceous and lower Tertiary clay layers at Gubbio, Italy. Geology 14:923-926.

Newmann, K.R.

1988      Palynomorph zones based on Cretaceous-Paleocene vertebrate fossil localities, McCone County, Montana. Geol. Soc. Amer. Abst. Prog. 20:459.

Nichols, D.J., D.M. Jarzen, C.J. Orth, and P.Q. Oliver

1986      Palynological and iridium anomalies at Cretaceous-Tertiary boundary, south-central Saskatchewan. Science 231:714-717.

Norton, N.J., and J.W. Hall

1969      Palynology of the Upper Cretaceous and lower Tertiary in the type locality of the Hell Creek Formation, Montana, U.S.A. Palaeontographica Abt. B, 125:1-64.

Novacek, M.J.

1986      The skull of leptictid insectivorans and the higher-level classification of eutherian mammals. Bull. Amer. Mus. Nat. Hist. 183:1-111.

Novacek, M.J., and W.A. Clemens

1977      Aspects of intrageneric variation and evolution of *Mesodma* (Multituberculata, Mammalia). Jour. Paleo. 51:701-717.

Officer, C.B., and C.L. Drake

1985      Terminal Cretaceous environmental events. Science 227:1161-1167.

Officer, C.B., A. Hallam, C.L. Drake, and J.D. Devine

1987      Late Cretaceous and paroxysmal Cretaceous/Tertiary extinctions. Nature 326:143-149.

Oltz, D.F., Jr.

1969      Numerical analyses of palynological data from Cretaceous and early Tertiary sediments in east-central Montana. Palaeontographica Abt. B, 128:90-166.

Orth, C.J., J.S. Gilmore, J.D. Knight, C.L. Pillmore, R.H. Tschudy, and J.E. Fassett

1981      An iridium abundance anomaly at the palynological Cretaceous-Tertiary boundary in northern New Mexico. Science 214:1341-1343.

1982      Iridium abundance measurements across the Cretaceous/Tertiary boundary in the San Juan and Raton basins of northern New Mexico, pp. 423-433 in L.T. Silver and P.H. Schultze, eds., Geological implications of impacts of large asteroids and comets on the earth. Geol. Soc. Amer. Spec. Paper 190.

Osborn, H.F.

1891      A review of Cretaceous Mammalia. Proc. Acad. Nat. Sci., Philadelphia, pp. 124-135.

1898      Evolution of the Amblypoda. Part 1. Taligrada and Pantodonta. Bull. Amer. Mus. Nat. Hist. 10:169-218.

Osborn, H.F., and C. Earle

1895      Fossil mammals of the Puerco beds. Collection of 1892. Bull. Amer. Mus. Nat. Hist. 7:1-70.

Parker, T.J., and W.A. Haswell

1897      A textbook of zoology: vol. 2. Macmillan and Co., London. 683 pp.

Prothero, D.R., E.M. Manning, and M. Fischer
1988    The phylogeny of the ungulates, pp. 236-239 in M.J. Benton, ed., The Phylogeny and Classification of the Tetrapods, vol. II. Syst. Assoc. Spec. Vol. 35B. Clarendon Press, Oxford.

Reig, O.A.
1981    Teoria del origen y desarrollo de la fauna de mamiferos de America del Sur. Monographiae Naturae, Publ. Mus. Municip. de Cienc. nat. "Lorenzo Scaglia." 162 pp.

Reig, O.A., J.A.W. Kirsch, and L.G. Marshall
1985    New conclusions on the relationships of opossum-like marsupials, with an annotated classification of the Didelimorphia. Ameghiniana 21:335-343.

Retallack, G.J.
1984    Completeness of the rock and fossil record: some estimates using fossil soils. Paleobiology 10:59-78.

Retallack, G.J., G.D. Leahy, and M.D. Spoon
1987    Evidence from paleosols for ecosystem changes across the Cretaceous/Tertiary boundary in eastern Montana. Geology 15:1090-1093.

Rigby, J.K., and J.K. Rigby, Jr.
1990    Geology of the Sand Arroyo and Bug Creek Quadrangles, McCone County, Montana. BYU Geology Studies 36:69-134.

Rigby, J.K., Jr.
1985    Paleocene dinosaurs—The reworked sample question. Geol. Soc. Amer. Abst. Prog. 17:262.
1987    The last of the North American Dinosaurs, pp. 119-153 in S. Czerkas and E. Olsen, eds., Dinosaurs Past and Present, vol. II. University of Washington Press, Seattle.
1989    The Cretaceous-Tertiary boundary of the Bug Creek Drainage: Hell Creek and Tullock formations, McCone and Garfield counties, Montana, pp. 67-73 in J.J. Flynn and M.C. McKenna, eds., Mesozoic/Cenozoic Vertebrate Paleontology: Classic Localities, Contemporary Approaches. Field Trip Guidebook T322, American Geophysical Union. Washington D.C.

Rigby, J.K., Jr., K.R. Newmann, J. Smit, S. Van der Kaars, R.E. Sloan, and J.K. Rigby
1987    Dinosaurs from the Paleocene part of the Hell Creek Formation, McCone County, Montana. Palaios 2:296-302.

Rigby, J.K., Jr., J.K. Rigby, and R.E. Sloan
1986    The potential for an unconformity near the Cretaceous/Tertiary boundary, basal Tullock Fm., McCone Co., MT. Geol. Soc. Amer. Abst. Prog. 18:730.

Rigby, J.K., Jr., and R.E. Sloan
1985    Dinosaur decline and eventual extinction near the Cretaceous-Tertiary boundary, Hell Creek Formation, Montana. Geol. Soc. Amer. Abst. Prog. 17:700.

Rogers, G.S., and W. Lee
1923    Geology of the Tullock Creek coal field, Rosebud and Big Horn counties, Montana. U.S. Geol. Surv. Bull. 749:1-181.

Russell, L.S.

1975    Mammalian faunal succession in the Cretaceous system of western North America, pp. 137-161 in W.G.E. Caldwell, ed., The Cretaceous System in the western interior of North America. Geol. Assoc. Can. Spec. Paper 13.

Savage, D.E.

1962    Cenozoic geochronology of the fossil mammals of the Western Hemisphere. Mus. Argent. Cienc. Nat., Cienc. Zool., Rev. 8:51-67.

Schmitz, B.

1988    Origin of microlayering in worldwide distributed Ir-rich marine Cretaceous/Tertiary boundary clays. Geology 16:1068-1072.

Scott, W.B.

1892    A revision of the North American Creodonta with notes on some genera which have been referred to that group. Proc. Acad. Nat. Sci., Philadelphia, 44:291-323.

Sholes, M.A., and G.A. Cole

1981    Depositional history and correlation problems of the Anderson-Dietz Coal Zone, southeastern Montana. Mountain Geol. 18:35-45.

Simmons, N.B., and M. Desui

1986    Paraphyly in *Catopsalis* (Mammalia, Multituberculata) and its biogeographic implications. Univ. Wyoming Cont. Geol. Spec. Paper 3:87-94.

Simpson, G.G.

1927a   Mammalian fauna of the Hell Creek Formation of Montana. Amer. Mus. Nov. 267:1-7.

1927b   Mesozoic Mammalia VIII: genera of Lance mammals other than multituberculates. Amer. Jour. Sci., ser. 5, 14:121-130.

1929a   American Mesozoic Mammalia. Mem. Peabody Mus. 3:1-171.

1929b   Some Cretaceous mammals from the Lance Formation. Ann. Carnegie Mus. 19:107-113.

1930    Post Mesozoic Marsupialia, pp. 1-87 in Fossilum Catalogus: 1. Animalia. Berlin.

1931    A new classification of mammals. Bull. Amer. Mus. Nat. Hist. 59:259-293.

Sloan, R.E.

1970    Cretaceous and Paleocene terrestrial communities of western North America, pp, 427-453 in E.L. Yochelson, ed., Proc. North Amer. Paleo. Conv. (1969). Allen Press, Kansas.

1976    The ecology of dinosaur extinction, pp. 134-154 in Athlon, Essays on Paleontology in Honor of Loris Russell. Misc. Publ. Royal Ontario Mus. Life Science.

1983    Late Cretaceous and Paleocene mammal ages, magnetozones, rates of sedimentation and evolution. Geol. Soc. Amer. Abst. Prog. 15:307.

1987    Paleocene and latest Cretaceous mammal ages, biozones, magnetozones, rates of sedimentation, and evolution, pp. 165-200 in J.E. Fassett and J.K. Rigby, Jr., eds., The Cretaceous-Tertiary Boundary in the San Juan and Raton Basins, New Mexico and Colorado. Geol. Soc. Amer. Spec. Paper 209.

Sloan, R.E., and J.K. Rigby, Jr.

1986    Cretaceous-Tertiary dinosaur extinction. Science 234:1170-1175.

Sloan, R.E., J.K. Rigby, Jr., L. Van Valen, and D. Gabriel
1986    Gradual dinosaur extinction and simultaneous ungulate radiation in the Hell Creek Formation. Science 232:629-633.

Sloan, R.E., and L. Van Valen
1965    Cretaceous mammals from Montana. Science 148:220-227.

Smit, J.
1985    Catastrophic events at the terrestrial Cretaceous-Tertiary boundary. Geol. Soc. Amer. Abst. Prog. 17:720.

Smit, J., and G. Klaver
1981    Sanidine spherules at the Cretaceous-Tertiary boundary indicate a large impact event. Nature 292:47-49.

Smit, J., and S. Van der Kaars
1984    Terminal Cretaceous extinctions in the Hell Creek area, Montana: Compatible with catastrophic extinction. Science 223:1177-1179.

Smit, J., S. Van der Kaars, and J.K. Rigby, Jr.
1987    Stratigraphic aspects of the Cretaceous-Tertiary boundary in the Bug Creek area of eastern Montana, U.S.A. Mem. de la Soc. Geol. de France, n.s., 150:53-73.

Storer, J.E.
1991    The mammals from the Gryde Local Fauna, Frenchman Formation (Maastrichtian: Lancian), Saskatchewan. Jour. Vert. Paleo. 11:350-369.

Sullivan, R.M.
1987    A reassessment of reptilian diversity across the Cretaceous-Tertiary boundary. Los Angeles Co. Mus. Cont. Sci. 391:1-26.

Szalay, F.S.
1982    A new appraisal of marsupial phylogeny and classification, pp. 621-640 in M.Archer, ed., Carnivorous Marsupials. Royal Society of New South Wales, Sydney.

Tedford, R.H.
1970    Principles and practices of mammalian geochronology in North America, pp. 666-703 in E.L. Yochelson, ed., Proc. North Amer. Paleo. Conv. (1969). Allen Press, Kansas.

Thom, W.T., and C.E. Dobbin
1924    Stratigraphy of Cretaceous-Eocene transition beds in eastern Montana and the Dakotas. Geol. Soc. Amer. Bull. 35:481-505.

Thomas, R.G., D.G. Smith, J.M. Wood, J. Visser, E.A. Calverley-Range, and E.H. Koster
1987    Inclined Heterolithic Stratification—Terminology, Description, Interpretation and Significance. Sedimentary Geology 53:123-179.

Tschudy, R.H.
1970    Palynology of the Cretaceous-Tertiary boundary in the Northern Rocky Mountain and Mississippi Embayment Regions. Geol. Soc. Amer. Spec. Paper 127:65-111.

Tschudy, R.H., C.L. Pillmore, C.J. Orth, J.S. Gilmore, and J.D. Knight
1984    Disruption of the terrigenous plant ecosystem after the Cretaceous-Tertiary boundary, western interior. Science 225:1030-1032.

Van Valen, L.

1966     Deltatheridia, a new order of mammals. Bull. Amer. Mus. Nat. Hist. 132:1-126.

1969     The multiple origins of the placental carnivores. Evolution 23:118-130.

1978     The beginning of the age of mammals. Evol. Theory 4:45-80.

1984     Catastrophes, expectations, and the evidence. Paleobiology 10:121-137.

Van Valen, L., and R.E. Sloan

1965     The earliest Primates. Science 150:743-745.

1977     Ecology and extinction of the dinosaurs. Evol. Theory 2:37-64.

Voight, E.

1981     Critical remarks on the discussion concerning the Cretaceous-Tertiary boundary. Newslet. Strat. 10:92-114.

Wilmarth, M.G.

1938     Lexicon of geologic names of the United States (including Alaska). U.S. Geol. Surv. Bull. 895:1-2396.

Winge, H.

1917     Udsigt over Insektaedernes indbyrdes Slaegtskab. Vidensk. Meddel. Dansk. Naturh. Foren. 68:83-203.

Wood, H.E., R.W. Chaney, J. Clark, E.H. Colbert, G.L. Jepsen, J.B. Reeside, Jr., and C. Stock

1941     Nomenclature and correlation of the North American continental Tertiary. Geol. Soc. Amer. Bull. 52:1-48.

Zoller, W.H., J.R. Parrington, and J.M.P. Kotra

1983     Iridium Enrichment in airborne particles from Kilauea Volcano: January 1983. Science 222:1118-1121.